Energy Materials

Energy Materials
A Short Introduction to Functional Materials for Energy Conversion and Storage

Aliaksandr S. Bandarenka

CRC Press is an imprint of the
Taylor & Francis Group, an **informa** business

First edition published 2022
by CRC Press
6000 Broken Sound Parkway NW, Suite 300, Boca Raton, FL 33487-2742

and by CRC Press
2 Park Square, Milton Park, Abingdon, Oxon, OX14 4RN

CRC Press is an imprint of Taylor & Francis Group, LLC

Library of Congress Cataloging-in-Publication Data
Names: Bandarenka, Aliaksandr S., author.
Title: Energy materials : a short introduction to functional materials
for energy conversion and storage / Aliaksandr S. Bandarenka.
Description: First edition. | Boca Raton, FL : CRC Press, 2022. |
Includes bibliographical references and index.
Identifiers: LCCN 2021037476 | ISBN 9780367457365 (hardback) |
ISBN 9780367458102 (paperback) | ISBN 9781003025498 (ebook)
Subjects: LCSH: Energy storage—Materials. | Fuel cells—Materials. |
Electric batteries—Materials. | Direct energy conversion—materials.
Classification: LCC TK2910 .B36 2022 | DDC 621.31/26—dc23/eng/20211110
LC record available at https://lccn.loc.gov/2021037476

ISBN: 978-0-367-45736-5 (hbk)
ISBN: 978-0-367-45810-2 (pbk)
ISBN: 978-1-003-02549-8 (ebk)

DOI: 10.1201/9781003025498

Typeset in Minion
by codeMantra

To my wife, Tamara, my parents, Liubou and Siarhei, and my children, Vera and Maxim.

Contents

Acknowledgment

I WANT TO THANK THE FOLLOWING COLLEAGUES for their valuable comments and suggestions regarding this manuscript: Dr. Batyr Garlyyev, Dr. Sebastian Watzele, Dr. Regina Kluge, Dr. Rohit Ranganathan Gaddam, Dr. Elena Gubanova, Leon Katzenmeier, Richard Haid, Xaver Lamprecht, Theophilus Kobina Sarpey, Thorsten Schmidt, and Leif Carstensen.

Last but not least, I want to thank my beloved spouse, Tamara Bandarenka, and my parents, Liubou and Siarhei Bandarenka, for their everyday support.

Preface

INTRODUCTION

Energy provision is among the main pillars of modern civilization. Nowadays, society experiences drastic changes in the paradigms and vision of how to provide energy sustainably, efficiently, and with a minimum negative impact on the environment. It is clear that the quality of life in urban and rural areas will fundamentally depend and already depends on the development of new and the evolution of existing energy provision schemes. Almost all developed countries recently declared that a transition from a fossil fuel economy to an economy based on renewable energy sources is one of the main priorities. Moreover, this transition already happens now with remarkable acceleration. However, to enable this process in the future and keep the current development level, a portfolio of ever-efficient energy materials is necessary. In this book, important classes of *functional energy materials* which are used today (for instance, heterogeneous catalysts for fossil fuel processing, permanent magnets for motors and generators, or semiconductors for solar cells), as well as materials which are under development for the near future (e.g., for the so-called artificial leaves), are considered. *Functional materials* are materials with given functionalities, which are of key significance for specific applications. Rather than dealing with the physical and chemical basics of energy conversion and storage, this work is addressed to a broader audience with the focus on various functional materials used in this field with simple explanations of their design principles, essential properties in terms of specific functionality, and quantitative figures of merit.

The topics related to modern energy materials are very interdisciplinary, requiring specific education in physics, chemistry, materials science, engineering, etc. However, this work is intended to be understandable to all students and scientists who work or start to work in these fields. It is also addressed to those who just want to have a quick and a broad first overview of the main concepts and materials used in energy materials.

The particular aims are as follows:

- To provide information on important state-of-the-art and promising functional energy materials at a level that readers with different backgrounds should understand
- To explain the origin of the functionality of those materials
- To explain the design principles of modern energy materials

One of the main hopes of this work is that it should also contribute to the "decarbonization" of the energy provision schemes.

Venite incipere!

Author

Professor Aliaksandr Bandarenka is currently a professor at the Technical University of Munich, Germany. He conducts research in the area of the physics of energy conversion and storage with the main research topics that include the design and implementation of functional materials for energy applications. The material design is based on various approaches using input from electrochemistry, solid-state physics, chemistry, and surface science and starts from model objects. Professor Bandarenka attended Belarusian State University. He earned his undergraduate degree in chemistry in 2002 and his PhD in 2005, working under the supervision of Dr. G.A. Ragoisha. After completing his PhD, he was a postdoctoral researcher at the University of Twente in the Netherlands. In this role, he worked with Prof. H.J.M. Bouwmeester and Prof. B.A. Boukamp on the development of new proton-conducting electrolytes. In 2008, he moved to the Technical University of Denmark where he worked with Prof. I. Chorkendorff and Prof. J. Nørskov on electrocatalysis for energy conversion. In 2010-2014, he was a group leader at the Center for Electrochemical Sciences (Director: Prof. W. Schuhmann) at the Ruhr-University Bochum, Germany. He is the recipient of the Materials Science Award (2013) from the International Society of Electrochemistry and the National Ernst Haage Prize (2016) from the Max Planck Institute for Chemical Energy Conversion (Germany).

CHAPTER 1

Energy and Fuels

1.1 WHAT ARE THE "BEST FUELS"?

Let's start the discussion in this chapter with a conventional definition of energy sources.

DEFINITION:

According to the United Nations *Concepts and Methods in Energy Statistics*, fossil fuels (like oil, coal, and natural gas) and "natural energy" (like hydropower, solar power, or even nuclear power) are collectively referred to as primary energy (sources). The term secondary energy (sources) is to designate all sources of energy that result from the transformation of primary sources.

We know that fuels are one of the main energy sources. There are various fuels nowadays, but choosing the best one depends on several application aspects and criteria. Let us first understand the requirements in terms of the volumetric and gravimetric energy density of fuels. This is probably the most straightforward approach: one wants to spend less (in terms of volumes and weights) and get the most (in terms of energy). With this in mind, one can look at Figure 1.1, where the volumetric and gravimetric energy densities are compared for the common fuels. It needs to be mentioned here that the best fuel should also allow the release of the maximum "useful" energy both per unit of volume and per unit of mass in relatively simple, cheap, and safe engines and reactors. It is clear that uranium (used in nuclear fission, nuclear reactors) wins in both nominations [1] with *ca* 1 500 000 GJ/L and *ca* 80 GJ/kg. However, several factors are limiting its usage now. First, the reactor design and necessary infrastructure are

DOI: 10.1201/9781003025498-1

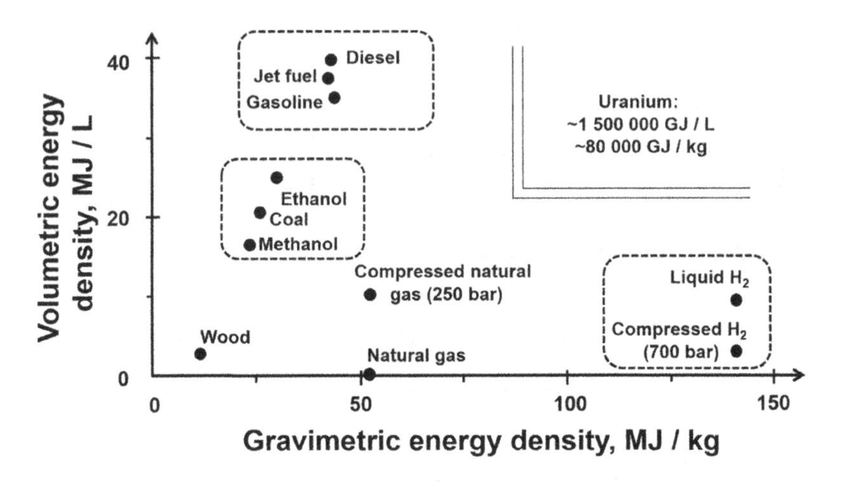

FIGURE 1.1 Volumetric vs. gravimetric energy densities for some standard fuels. Note that the exact values usually depend on several parameters in each particular case, i.e., purity, type of the reactor, etc.

complicated and expensive to implement in practice. Worldwide, to build a 1,200 MW reactor, one needs up to 10 years, with the entire infrastructure, which costs from 6 to 10 billion US dollars [2]. Another complication is related to the safety issues, problems with storing the radioactive wastes, and disassembling of the reactors after their use [3]. This is why fossil fuels are widely in use today, as they are the next alternatives (Figure 1.1). Diesel, jet fuels, and gasoline can be produced from the available oil. The former are liquids, and this fact simplifies their transportation and consumption in conventional combustion engines. However, the energy applications of oil are commonly not safe for human health and the environment [4] due to the generation of cancer-causing pollutants, toxic CO, and greenhouse gases like CO_2, which are released into the atmosphere after their burning [5,6]. The same environmental and health problems are common in energy applications of coal or natural gas [7].

Taking into account the concerns mentioned above, various participants of the global and local energy markets have begun considering hydrogen (H_2) as a viable fuel in everyday applications [8–12]. It has a relatively large gravimetric energy density and quite good volumetric energy density, whether compressed or liquefied, as shown in Figure 1.1. Moreover, the main ecological advantage of H_2 used as a fuel is that it leaves only pure water as an exhaust – a perfect solution for the urbanized areas and the environment in general. Unfortunately, there are no natural sources of pure gaseous hydrogen on the Earth: it needs to be procured from resources like water or natural gas. This requires additional energy, and many existing technologies and infrastructures are not ready for the broader commercialization of this fuel. These issues will be considered later in this chapter.

Accounting for the concerns mentioned earlier, it is not easy to answer the question on the best fuel: the problem is multiparametric, and it looks like there are no simple ideal solutions for the nearest future. The challenge is even more complicated, as one should consider that the existing infrastructure is created for fossil fuel usage, with massive investments already made. This is also clear from the graph showing the current contribution of the available fuels and energy technologies to the annual world energy consumption, which was $\sim 6 \times 10^{20}$ joules in 2019 (see Figure 1.2).

The figure demonstrates that the energy market is mainly dependent on oil, natural gas, and coal. Nuclear technologies can be considered as "outsiders" in this respect. In Figure 1.2, hydro sources and the so-called "renewables" (other than hydro energy) are also indicated. What are those?

DEFINITION:

Renewable energy sources (renewables) are natural energy sources that are inexhaustible from the viewpoint of the current human standards. These include solar, biomass, hydroelectricity, wind energy, ambient heat, geothermal, and ocean energy.

According to conservative forecasts, fossil fuels will likely continue to supply almost 80% of the world's energy consumption by 2040. However,

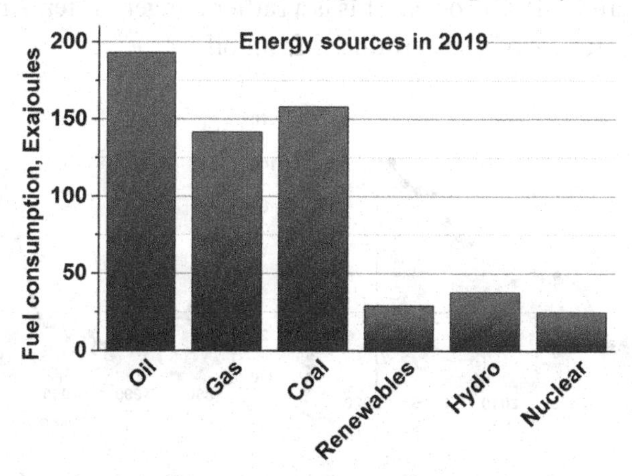

FIGURE 1.2 Annual world energy consumption by fuels and energy sources according to [13].

renewables are the world's fastest-growing energy sources that increased by *ca* 3.2 EJ in 2019 (among them, wind energy contributed ~1.4 EJ and solar energy ~1.2 EJ) and demonstrated ~11% growth per year [14]. With the rapid depletion of fossil fuel resources, one could envisage even faster growth in renewables in the near future [15,16].

Probably, one of the very sound concerns about the availability and the future of fossil fuels for the global economy and energetics was that stated in the influential 1998 Scientific American article entitled *The End of Cheap Oil*. The authors claimed that the "global production of conventional oil would begin to decline sooner than most people think, probably within 10 years" [17]. These lines of thought led to an upsurge in oil prices and contributed to the energy crisis around 2007–2008. However, as one can see from Figure 1.3a, global oil production was in continuous growth. The amount of oil produced annually in the present time is at an all-time high, with ~95 million barrels produced daily. Figure 1.3b, in turn, shows large fluctuations in the average oil prices for the last four decades. These large price fluctuations destabilize the world's economy, stimulating the search for alternative sustainable energy solutions.

As one can see from Figure 1.4, the annual discoveries of natural oil deposits were maximal between 1960 and 1970, when the leading market players could discover almost 60 billion barrels of oil per year. After that, even relatively high oil prices could not invert the tendency of declining discoveries. Today, the companies worldwide discover only one barrel of oil for every seven the society consumes globally (compare the data in Figures 1.3a and 1.4). Of course, this is a rather dangerous tendency, which questions the future of fuels produced from oil.

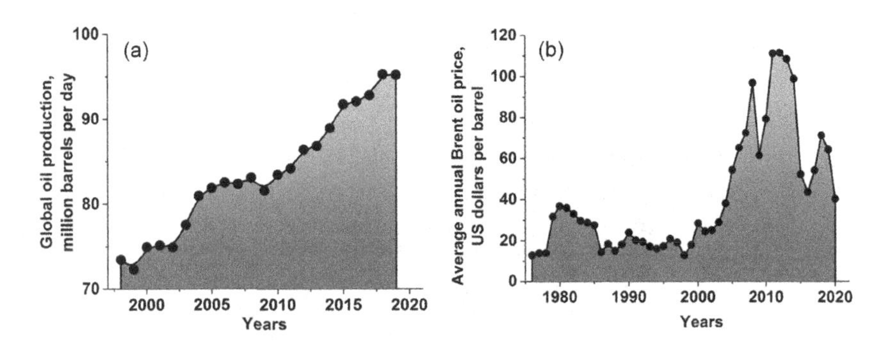

FIGURE 1.3 (a) Global annual oil production in 1998–2019 and (b) average Brent oil prices according to [13,14]. The annual consumption of oil in 2019 was ~35 billion barrels.

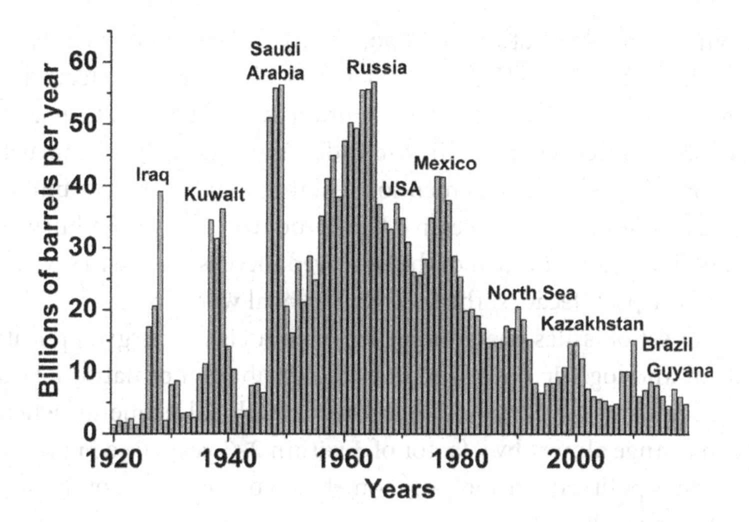

FIGURE 1.4 Global annual oil discoveries in 1920–2019, based on [26–28]. The names of the regions on the graph indicate the significant oil discoveries during the corresponding periods.

Each country would probably solve the energy problems of the future differently, depending on the available resources and dominating visions [18–25]. Some countries or even regions will continue to focus on fossil fuels. Some will continue to consider nuclear reactors as the pillars of the electricity provision, and some will widely rely on hydropower. However, there is a growing awareness and understanding that one needs new alternative scenarios and concepts, like the so-called hydrogen economy discussed below.

1.2 THE CONCEPT OF THE HYDROGEN ECONOMY

As mentioned above, there are several good rationales for reconsidering energy provision and storage paradigms today. The first rationale deals with the quality of life of the people. Nowadays, one can observe the next loop of urbanization: more than 50% of the world's population lives in urban areas, and almost all the countries are becoming increasingly urbanized [29]. Today, the forecast is that more than 70% of the planet's population will live in urban areas by 2050 [30]. The everyday life of these people will be primarily affected by the energy provision schemes to be used. The urban energy systems and industry [31] are currently based mainly on fossil fuels; this causes serious ecological problems.

The exhausts of the combustion engines of cars and the power plants predominantly poison the air in the cities. It has been recently estimated that environmental pollution (total pollution) kills more people every year than smoking, hunger, AIDS, tobacco and alcohol use, road accidents, or regional conflicts (Figure 1.5). Air pollution is the "top health hazard in Europe" [32], which escalates in the industrially developed and urbanized regions. One should also mention a higher than 90% probability that greenhouse gases such as excessive CO_2 produced due to the burning of fossil fuels have caused much of the observed increase in Earth's temperatures over the past decades, the so-called "global warming".

Another set of issues related to using fossil fuels in energy applications is political and logistic ones. The uneven distribution of natural carbon-based energy sources largely destabilizes the global economy when oil price can change almost by a factor of 3 within 2–3 years (see Figure 1.3b). It also creates political tensions and sometimes becomes one of the significant factors provoking local wars.

Finally, one should recall the famous Mendeleyev's opinion about petroleum expressed *ca* 150 years ago: "To burn oil means to stoke a stove

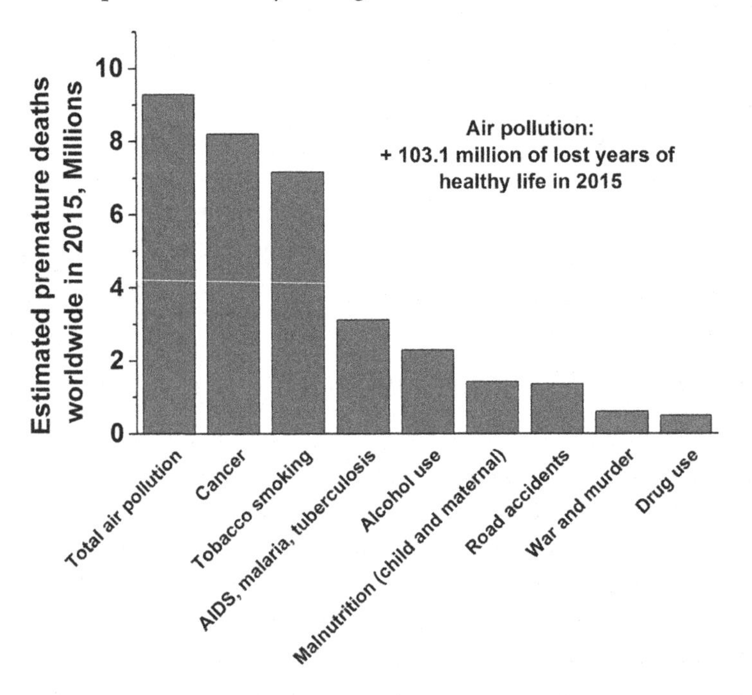

FIGURE 1.5 Global estimated premature deaths in the world in 2015. The data are from [33–35]. Naturally, there are differences in the methodologies and ways to distinguish, e.g., the cancer cases. However, the picture gives the impression that air pollution is among the main problems nowadays.

with banknotes". It means that oil is not only a source of fuels but a very precious and nonrecoverable source of chemicals for numerous industrial processes. Its use is versatile and not only limited to combustion engines.

Ecological problems and depletion of fossil fuel sources call for carbon-neutral or, ideally, carbon-free energy provision technologies. These technologies can use either "solar fuels" [36,37], approaches of "hydrogen economy", or hybrid energy schemes. For instance, there is a long-lasting dream to move toward electric vehicles in automotive applications, which use hydrogen as an initial fuel or are powered by electricity generated using renewable energy sources. Interestingly, it was the electric car, which for the first time reached a speed of over 100 km/h ("La Jamais Contente") in 1899. Nowadays, the interest in such autos is enormously renewed, and more and more commercial electric cars, both battery- and hydrogen-based, are on the roads.

DEFINITION:

The term *hydrogen economy* refers to a vision of using hydrogen as a carbon-free or low-carbon energy source replacing, for example, gasoline as a transport fuel or natural gas as a heating fuel (after John Bockris's talk in 1970) [38]. The key idea is that zero carbon emission renewable energy sources should be involved to produce H_2 from water using wind power, solar power, hydropower, wave power, or similar.

It would be worth noting that in the present scenario, only ~4% out of all hydrogen produced in the world annually (~60 Mt [7] were in total produced in 2017) is generated by water electrolysis, while the rest comes mainly from natural gas and other hydrocarbon sources [39]. Further, electricity to perform such electrolysis is currently generated primarily using various fossil fuels. An alternative vision is to use the same zero-emission renewables and produce liquid fuels like methanol or ethanol using atmospheric CO_2 and H_2O (zero- or low-carbon emission cycles). Such liquid "solar fuels" are also very attractive options: It is relatively easy to store and transport them. They have a high energy density, and one can use the already existing infrastructure.

Hence, the idea seems to be very simple: let's use the electricity generated, for instance, by wind power, solar power, or hydropower and

i. Perform water electrolysis in the absence or in the presence of atmospheric CO_2 to produce H_2 fuel or "solar fuels" to store energy practically in chemical bonds

ii. Store the thus generated electrical energy in specific storage devices like batteries, supercapacitors, or superconductor storage devices

iii. Use hydrogen or "solar fuels" when necessary using special devices called fuel cells or even conventional combustion engines

When necessary, one can even use conventional fossil fuels in some cases, which will not then contribute much to the energy balance and, therefore, would not significantly influence the environment.

To make this happen efficiently, one should solve the so-called *generation vs. consumption problem* (which is also essential nowadays in traditional energy provision schemes) at a new level. For instance, the generation of electricity and its consumption are typically separated in time: the wind is usually not constant, and the sun is not always shining. At the same time, electricity consumption is very different during the day and at night. One should fill this gap by storing electrical energy using special storage devices or in chemical bonds. That would result in many challenges, and with the ~170 GW of storage capacity worldwide existing in 2018 (more than 96% of which is provided by pumped hydro systems) [40], it would be challenging to address them quickly.

One can generate excessive amounts of hydrogen, but how can one store it efficiently until used? Suppose one wants to generate hydrogen and solar fuels through electrolysis at low costs [41] or consume them efficiently. In that case, one needs substances called catalysts and electrocatalysts to accelerate their production and transformation per unit of time. In the case of electrolysis, suitable liquid and solid electrolytes are necessary to reduce associated resistive losses. It should be noted here that there is still a certain collective belief that electricity from wind, sun, *etc.*, "is almost for free". However, to make use of wind energy, one must have special generators. For example, generators in wind turbines require affordable permanent magnets, which should be highly resistant to demagnetization. Those permanent magnets are also necessary for the electric vehicles used in automotive applications. To convert energy from direct sunlight, the use of solar cells is indispensable [42,43]. The latter require semiconductors and transparent electron conductors as the key materials. Finally, the generated electricity should be stored for portable, automotive, and stationary large-scale applications. For that, one needs electrode materials for different types of batteries, supercapacitors, superconducting storage systems, etc. The critical materials listed above can be called *functional energy materials*, and those will be considered in the following, step by step, in more detail. A particular focus will be set on their properties, understanding, design principles, and applications in energy provision schemes.

1.3 SUMMARY AND CONCLUSIONS

Sustainable energy provision and storage have become a more and more critical issue due to population growth and the development of new technologies. The current usage rate of fossil fuels considerably questions a dynamic and safe future development of energy systems for the growing world economy. To address the rising concerns, better *functional energy materials* for safe and viable energy provision and storage are necessary, and a clear understanding of the physics of energy conversion is essential [44]. There is a growing awareness currently that the possible strategies to increase such sustainability and viability include the optimizations of traditional technologies of the usage of fossil fuels, the development of the hydrogen economy and/or "solar fuel" production, as well as, in principle, the application of combined approaches.

1.4 QUESTIONS

1. Define *primary* and *secondary* energy sources.
2. What is the global energy consumption by source, and what are the renewable energy sources?
3. What is the energy density of the common fuels? Analyze the problems and alternatives of the current fossil fuel economy.
4. What is the concept of *hydrogen economy*?
5. What are the possible ways to address the electricity *generation versus consumption* problem?
6. What are *solar fuels*?
7. The role of material science in energy provision schemes: what kind of materials is necessary?

REFERENCES

1. Uranium for nuclear power: Resources, mining and transformation to fuel (Ed: Hore-Lacy, I.). 2016. Elsevier.
2. Walls, J. 2011. Nuclear power generation – past, present and future. In: *Nuclear Power and the Environment*. Editors: Harrison, R.M.; Hester, R.E., John Wiley & Sons: New York, pp. 1–39.
3. Smith, J.T. 2011. Nuclear accidents. In: *Nuclear Power and the Environment*. Editors: Harrison, R.M.; Hester, R.E., John Wiley & Sons: New York, pp. 57–81.
4. Puertas, R.; Marti, L. 2021. International ranking of climate change action: An analysis using the indicators from the Climate Change Performance Index. *Renewable and Sustainable Energy Reviews* 148:111316.
5. Uddin, M.N.; Wan Daud, W.M.A. 2014. Technological diversity and economics: Coupling effects on hydrogen production from biomass. *Energy and Fuels* 28:4300–4320.
6. Won, W.; Kwon, H.; Han, J.H.; Kim, J. 2017. Design and operation of renewable energy sources based hydrogen supply system: Technology integration and optimization. *Renewable Energy* 103:226–238.
7. Parkinson, B.; Balcombe, P.; Speirs, J.F.; Hawkes A.D.; Hellgardt. K. 2019. Levelized cost of CO_2 mitigation from hydrogen production routes. *Energy and Environmental Science* 12:19–40.
8. International Energy Agency. 2019. The future of hydrogen: seizing today's opportunities. Report prepared by the IEA for the G20, Japan.
9. Dawood, F.; Anda, M.; Shafiullah, G.M. 2020. Hydrogen production for energy: an overview. *International Journal of Hydrogen Energy* 45: 3847–3869.
10. Abdallaa, A.M.; Hossaina, S.; Nisfindya, O.B.; Azadd, A.T.; Dawoodb, M.; Azad, A.K. 2018. Hydrogen production, storage, transportation and key challenges with applications: a review. *International Journal of Hydrogen Energy* 165:602–627.
11. Veras, T.S.; Mozer, T.S.; dos Santos, D.C.R.M.; César, A.S. 2017. Hydrogen: trends, production and characterization of the main process worldwide. *International Journal of Hydrogen Energy* 42: 2018–2033.
12. H2.live, Hydrogen stations in Germany and Europe. 2021. https://h2.live/en (Accessed June 2021).
13. Fuel shares of primary energy and contributions to growth in 2019. 2020. *BP Statistical Review of World Energy* 69:16.
14. Fuel shares of primary energy and contributions to growth in 2019. *BP Statistical Review of World Energy* 69:3.
15. Doman, L. 2017. EIA projects 28% increase in world energy use by 2040, U.S. Energy Information Administration (EIA). *Today in Energy*. https://www.eia.gov/todayinenergy/detail.php?id=32912 (Accessed: June 2021).
16. Stram, B.N. 2016. Key challenges to expanding renewable energy. *Energy Policy* 96:728–734.
17. Campbell, C.J.; Laherrère, J.H. 1998. The end of cheap oil. *Scientific American* March:78–83.

18. International Energy Agency. World Energy Outlook 2019. 2019. Executive summary, Paris.
19. Østergaard, P.A.; Duic, N.; Noorollahi, Y.; Mikulcic, H.; Kalogirou, S. 2020. Sustainable development using renewable energy technology. *Renewable Energy* 146:2430–2437.
20. Bhandari, R.; Shah, R.R. 2021. Hydrogen as energy carrier: Techno-economic assessment of decentralized hydrogen production in Germany. *Renewable Energy* 177:915–931.
21. Destek, M.A.; Sinha, A. 2020. Renewable, non-renewable energy consumption, economic growth, trade openness and ecological footprint: Evidence from organisation for economic co-operation and development countries. *Journal of Cleaner Production* 242:118537.
22. Cheng, C.; Blakers, A.; Stocks, M.; Lu, B. 2019. Pumped hydro energy storage and 100% renewable electricity for East Asia. *Global Energy Interconnection* 2:386–392.
23. Taie, Z.; Villaverde, G.; Morris, J.S.; Lavrich, Z.; Chittum, A.; White, K.; Hagen, C. 2021. Hydrogen for heat: Using underground hydrogen storage for seasonal energy shifting in northern climates. *International Journal of Hydrogen Energy* 46:3365–3378.
24. Kakoulaki, G.; Kougias, I.; Taylor, N.; Dolci, F.; Moya, J.; Jäger-Waldau, A. 2021. Green hydrogen in Europe – a regional assessment: Substituting existing production with electrolysis powered by renewables. *Energy Conversion and Management* 228:113649.
25. Schlachtberger, D.P.; Brown, T.; Schramm, S.; Greiner, M. 2017. The benefits of cooperation in a highly renewable European electricity network. *Energy* 134:469–481.
26. Rystad Energy ECube, January 2020 and Rystad Energy research and analysis 2020. https://www.rystadenergy.com/newsevents/news/press-releases/global-oil-and-gas-discoveries-reach-four-year-high-in-2019/. (Accessed: November 2020).
27. Ganser, D., 2011. Peak oil und die clean-tech chance der Schweiz. Swiss Institute for Peace and Energy Research, Basel, 11 p.
28. Zittel, W.; Weindorf, W. 2010. Erneuerbare Energien und Energieeffizienz als zentraler Beitrag zur Europäischen Energiesicherheit; Teilstudie Reserven und Fördermöglichkeiten von Uran bis 2050, Studie im Auftrag des Bundes ministeriums für Umwelt, Naturschutz und Reaktorsicherheit. www.bmu.de/files/pdfs/allgemein/application/pdf/um08__41_819_bf.pdf (Accessed: November 2020).
29. United Nations. 2020. Department of economic and social affairs, population division. https://www.un.org/en/development/desa/population/theme/urbanization/index.asp (Accessed: October 2020).
30. United Nations. 2019. Department of economic and social affairs, population division. *World urbanization prospects: The 2018 revision (ST/ESA/SER.A/420)*. New York: United Nations.
31. Thiel, G.P.; Stark, A.K. 2021. To decarbonize industry, we must decarbonize heat. *Joule* 5:531–550.

32. Air pollution is 'top health hazard in Europe'. 2016. DW. https://p.dw.com/p/2T6bD (Accessed: October 2020).
33. Tong, S. 2019. Air pollution and disease burden. *The Lancet Planetary Health* 3:E49–E50.
34. Pollution killing more people than war and violence. 2017. DW. https://p.dw.com/p/2mDUT (Accessed: October 2020).
35. Cohen, A.J. et al. 2017. Estimates and 25-year trends of the global burden of disease attributable to ambient air pollution: an analysis of data from the Global Burden of Diseases Study 2015. *Lancet* 389:1907–1918.
36. Shih, C.F.; Zhang, T.; Li, J.; Bai, C. 2018. Powering the future with liquid sunshine. *Joule* 2:1925–1949.
37. Montoya, J.H.; Seitz, L.C.; Chakthranont, P.; Vojvodic, A.; Jaramillo, T.F.; Nørskov, J.K. 2017. Materials for solar fuels and chemicals. *Nature Materials* 16:70–81.
38. Bockris, J.O.M. 2013. The hydrogen economy: Its history. *International Journal of Hydrogen Energy* 38:2579–2588.
39. Fakeeha, A.H.; Ibrahim, A.A.; Khan, W.U.; Seshan, K.; Al Otaibi, R.L.; Al-Fatesh, A.S. 2018. Hydrogen production via catalytic methane decomposition over alumina supported iron catalyst. *Arabian Journal of Chemistry* 11:405–414.
40. Gür, T.M. 2018. Review of electrical energy storage technologies, materials and systems: challenges and prospects for large-scale grid storage. *Energy & Environmental Science* 11:2696–2767.
41. Vesborg, P.C.K.; Jaramillo, T.F. 2012. Addressing the terawatt challenge: scalability in the supply of chemical elements for renewable energy. *RSC Advances* 2:7933–7947.
42. Victoria, M.; Haegel, N.; Peters, I.M.; Sinton, R.; Jäger-Waldau, A.; del Cañizo, C.; Breyer, C.; Stocks, M.; Blakers, A.; Kaizuka, I.; Komoto, K.; Smets, A. 2021. Solar photovoltaics is ready to power a sustainable future. *Joule* 5:1041–1056.
43. Jiang, M.; Li, J.; Wei, W.; Miao, J.; Zhang, P.; Qian, H.; Liu, J.; Yan, J. Using existing infrastructure to realize low-cost and flexible photovoltaic power generation in areas with high-power demand in China. *iScience* 23:101867.
44. Krischer, K.; Schönleber, K. 2015. *Physics of Energy Conversion*. De Gruyter: Berlin.

CHAPTER 2

Heterogeneous Catalysts for Fuel Processing

2.1 CATALYSIS IN ENERGY APPLICATIONS

As discussed in the previous chapter, the market of fossil fuels is currently distributed between oil (~12 Mt consumed daily), coal (~22 Mt per day), and natural gas (up to ~10,000 Mm^3 of daily consumption) [1]. However, it is essential to use them rationally, as they are not only fuels but at the same time valuable sources of precursors for the chemical industry to produce goods of everyday importance like plastics, lubricants, etc. [2]. Considering this fact and environmental issues, one can observe modern tendencies, which aim to limit the use of fossil fuels in energy applications. Of course, one straightforward way to do so is to optimize or adapt existing technologies to minimize the consumption of these resources continuously. However, even for that, the industry currently needs new efficient technologies and functional materials to keep and improve the current quality of life worldwide.

The rational schemes of how to optimize the use of fossil fuels include multiple scenarios [3]. Several examples are given below.

First of all, relatively affordable natural gas (mainly CH_4) can be transformed into liquid fuels that are particularly important for automotive applications, according to the following scheme:

$$CH_4 + H_2O \rightarrow CO + 3H_2 \qquad \text{Steam(methane)reforming}$$

$$(2n+1)H_2 + nCO \rightarrow C_nH_{(2n+2)} + nH_2O \qquad \text{Fischer-Tropsch process}$$

DOI: 10.1201/9781003025498-2

One can also recall a well-known process of converting coal into synthetic fuels, both liquid and gaseous, like methane, CH_4 (Bergius process) [4,5]:

$$nC+(n+1)H_2 \rightarrow C_nH_{2n+2}$$

The schemes mentioned above can also be used to address the energy generation vs. consumption problems. For instance, the energy from the nuclear power plants working during the night, when the energy consumption is low, can be temporarily transformed into carbon-free (e.g., H_2) or carbon-containing fuels [6–8]. The latter can be afterward used to obtain heat or electricity during the day when the consumption is high.

Another set of approaches assumes that low-quality oil and heavy hydrocarbons, which are often byproducts of oil processing, can be transformed with a great added value to the high-quality liquid fuels, for instance, jet fuels [9,10]. Those are the middle distillate products with a low content of aromatic compounds and a boiling point between 150°C and 290°C. Commercial and widely available technology for this within the existing industrial schemes is *hydrocracking* [11–13]. The goal of this process is to break long-chain hydrocarbons into short-chain hydrocarbons (see Figure 2.1). Hydrocracking consumes less thermal energy and is more practical and selective compared to the other related alternative schemes popular so far. One can potentially use some waste organic substances, *e.g.*, vegetable oils or biomass [14,15], to produce standard fuels

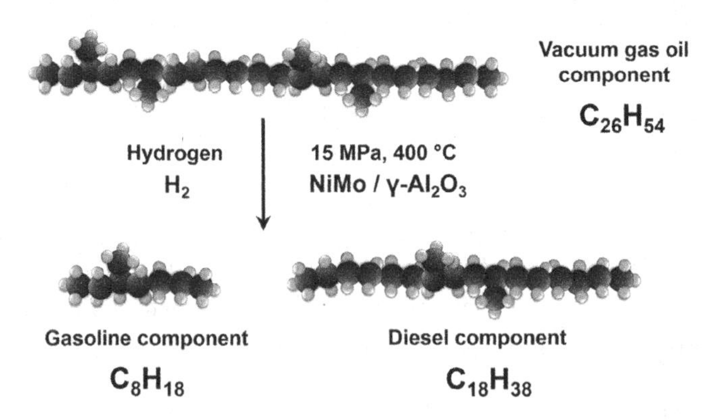

FIGURE 2.1 Example of a hydrocracking process converting heavy oils (e.g., vacuum gas oil), which are leftovers from petroleum distillation, to gasoline and Diesel fuels. The darker colors refer to the carbon atoms. The common catalysts are Al_2O_3-supported Ni/Mo-based catalysts.

or additives [16]. Nowadays, hydrocracking units are among the most commercially profitable ones in many companies.

The efficiency of converting carbon or natural gas to high-quality liquid fuels or the performance of the hydrocracking units is fundamentally dependent on the functional energy materials, which are called *catalysts* [17,18]. The common definition of these materials and the phenomenon of catalysis are given below.

DEFINITION:

Catalysis is a phenomenon by which chemical reactions are accelerated by relatively small quantities of substances, called catalysts. The definition proposed by Wilhelm Ostwald (Nobel Prize winner 1909), using the approaches of the kinetics of chemical reactions, states that "a catalyst is a substance influencing the rate of a reaction; however, it is not a part of the products" [19]. It is important to realize that an appropriate catalyst only enhances the rate of a thermodynamically possible reaction. It cannot change the position of the thermodynamic equilibrium.

In turn, *heterogeneous catalysis* involves systems in which the catalyst and reactants form separate physical phases [20]. Understanding and designing heterogeneous catalysts are of paramount importance [21,22], as they are key materials not only for energy applications. Close to 80% of industrial chemical processes use them. It is due to the relative simplicity of extracting the products after the reactions.

Common catalytic reactions proceed from reactants to products through one or several species called intermediates. This gives a basis to introduce the so-called *reaction coordinate*, which defines the route from the reactants to the products. Probably one of the simplest and most widely used energy diagrams to illustrate the effect of the catalysts is shown in Figure 2.2 for the case of a reaction with only one type of reaction intermediates adsorbed at the catalyst surface (an example with simple schemes is shown below the corresponding axis). As one can see, the energy barrier on the way from the reactant(s) to the product(s) is much lower in the case of a suitable catalytic surface. From this model, it is intuitively understandable that one can expect a higher probability to jump from the "reactant position" to the "product position" for a lower barrier. This is how in simple qualitative terms one can explain how pure energetic considerations may explain kinetic phenomena. However, this scheme only describes the overall concept of catalysis. It does not say anything about the multistep reactions (with various intermediates) or how to design the actual catalysts.

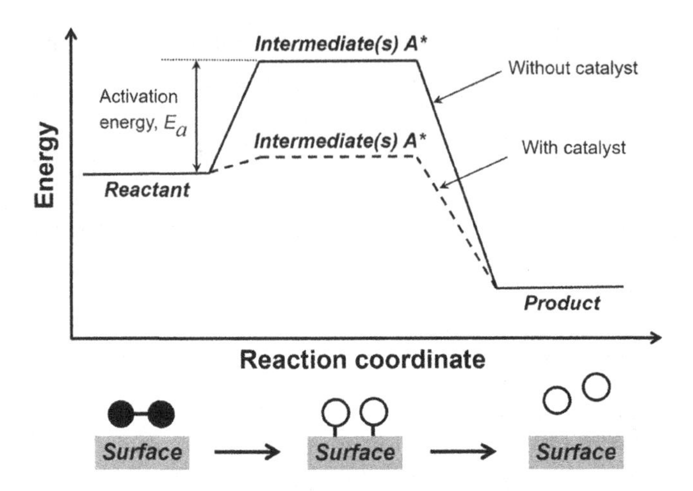

FIGURE 2.2 Schematic energy diagram for a catalytic reaction at a surface: Reactant → Intermediate(s) A* → Product. The activation energy, E_a, to form the intermediates is much lower if the surface exhibits the catalytic properties. Note that the reaction coordinate just schematically describes the way from the reactant to the product.

2.2 CATALYST ACTIVITY

Attempts to understand how heterogeneous catalysts work have a relatively long history. Probably the first step was the idea proposed by P. Sabatier in *ca* 1911 [23]. It is a qualitative concept postulating that to maximize the activity, i.e., to increase the amount of product per unit of time for a given amount of a catalyst, the interactions between the catalyst and the reaction intermediates should be just right. The catalyst surface should bind the intermediates neither too strong nor too weak. If the interaction is too weak, the reactant will fail to bind to the catalyst, and no reaction will take place. On the other hand, if the interaction is too strong, the catalyst gets blocked (poisoned) by the intermediates. Here, one can have a rough analogy with situations met in some fast-food restaurants. The seats in such restaurants are often designed comfortable enough to attract clients, but at the same time, they are slightly uncomfortable to persuade people to leave the place soon after finishing their meal, thus maximizing the number of persons per unit of time and per seat (in case of catalysts – "turnover frequency", the amount of products per unit of time for a given part of a catalytic surface) bringing the maximal income to the restaurant.

After P. Sabatier, Irvine Langmuir (Nobel Prize winner 1932) postulated in 1922 that "in general, a catalyst surface is quite complex, and it resembles a checkerboard in which some of the spaces are vacant, while others are filled with atoms or molecules" [24]. Later, in 1925, Hugh Stott Taylor suggested that "a catalyzed chemical reaction is, e.g., not catalyzed over the entire solid surface of the catalyst but only at certain *active sites* or centers" [25]. This hypothesis is now considered as the first introduction of the concept of active catalytic sites, at which certain chemical reaction is fast. Namely, these sites at the surface should bind the intermediates just right. According to such a concept, the active centers can be located at atoms with different coordination and chemical nature.

In the second half of the 20th century, numerous attempts to "quantify" the Sabatier principle resulted in the so-called *volcano-plot* approach [26]. The volcano plots correlate the activity (the rate of a catalytic chemical reaction) with a descriptor (describing the stability of the relevant reaction intermediates at the surface) [27]. The descriptor can be the heat of adsorption [28] of one of the reactants or the heat of formation of a bulk compound relative to the surface compound, or simply the position of the catalytic material (e.g., metal) in the periodic table of elements. However, it is important to find such a parameter, which is easy to measure or assess.

Directly from the Sabatier principle and the Tailor's concept of active sites, the suitable descriptor is the heat of adsorption or *binding energy* of the reaction intermediates to the catalytic centers. In Figure 2.3, a schematic volcano plot is given. As one can see from the Figure, if a suitable descriptor is selected, there is its specific optimum value at which the activity is maximal. With the volcano plots, one can rank different

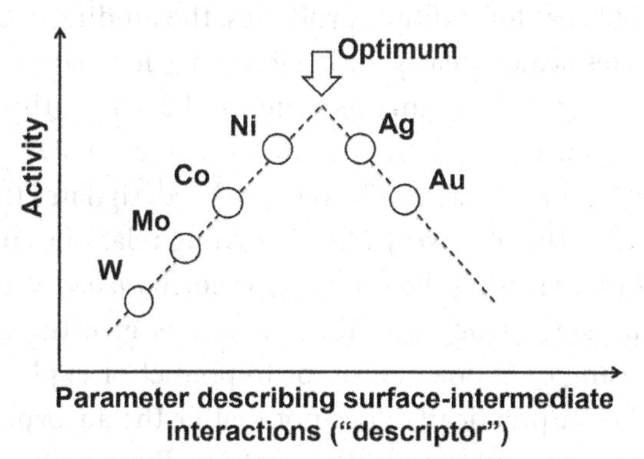

FIGURE 2.3 A scheme of a volcano plot used in heterogeneous catalysis to quantify the Sabatier principle. It is essential to select the descriptor, which appropriately describes the strength of the interactions between the surface and intermediates.

catalytic surfaces with respect to their potential for further optimizations and modifications for real-world applications.

However, there are some challenges when applying such an approach to analyze the catalytic processes. The first question arises for multistage reactions. Indeed, if multiple dissimilar intermediates are involved, the binding energy of which of them should be used as the descriptor? Another question is how to assess the descriptor quickly and affordably. Experimental measurements of the heat of adsorption are generally very time-consuming and demanding. The predicting power of such an approach is questionable as the experimental assessment of the descriptor is practically more complicated than the activity measurements themselves. Therefore, the use of measured descriptors is limited; they are often used to *explain* the observed activity trends, not to *predict* them. Recently, with the development of theoretical methods and computational power, it became possible to calculate, *e.g.*, the binding energies with acceptable accuracy by utilizing density functional theory (DFT) calculations [29]. In that approach, one can analyze the adsorption properties of known catalytic objects and also hypothetical surfaces. Theoretical analysis of binding energies has become an excellent alternative to the experimentally measured descriptors, improving significantly the predictive power of the Sabatier approach.

A noticeable breakthrough in understanding the activity trends for the multistage catalytic reactions was achieved in *ca* 2007 when a series of theoretical works showed the existence of the so-called *scaling relations* [30]. It turned out that for multiple processes, the binding energies of reaction intermediates scale linearly with each other for many surfaces if the mechanism or the reaction remains roughly the same. This is schematically shown in Figure 2.4 for the reaction scheme involving three intermediates, A*, B*, and C*, where the symbol "*" designates the adsorption site (active center). The discovery of such scaling relations enabled certain simplifications in analyzing the catalytic performance of various surfaces toward the multistage reactions. These relations give the clear physical rationale that using only one descriptor to predict or explain the activity trends is possible. It practically does not matter the adsorption energy of which intermediate is involved in the analysis. For example, for the case shown in Figure 2.4, one can use just one of the binding energies as the descriptor: ΔE_{A^*}, ΔE_{B^*}, or ΔE_{C^*}. The relative activity ranking of the surfaces in the volcano plots will remain approximately the same; only the values in the descriptor axes will change.

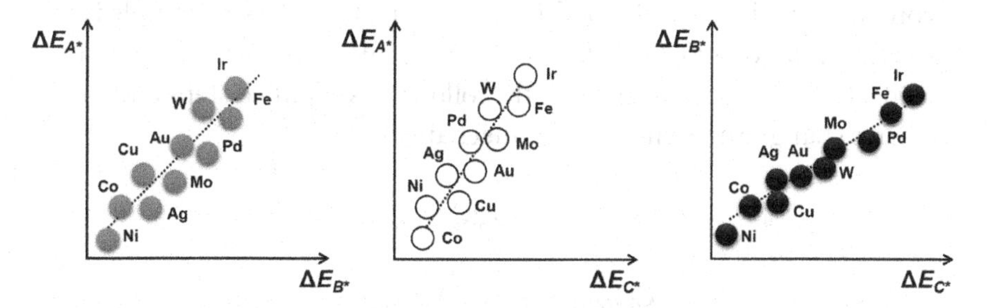

FIGURE 2.4 Schematic representation of the scaling relations for a multistage catalytic reaction. The binding energies for reaction intermediates A*, B*, and C* scale approximately linearly with each other for different catalytic surfaces, as long as the reaction mechanism remains the same.

Now, an interesting question can be asked: what is the physical origin of the scaling relations? The answer to this question is still under debate. However, a rough approximation, which can be used to explain a majority of known cases, is the following. The scaling relations will likely be observed if all the reaction intermediates of a multistage catalytic reaction are adsorbed at the active site through the atoms of the same chemical nature. One can illustrate this using a scheme shown in Figure 2.5 for the Fischer-Tropsch synthesis of fuels from the gas mixture of CO and H_2 [31].

In the Figure, one can see that irrespective of the resulting length of the chain of the growing hydrocarbon molecule, it is always adsorbed through the atom of the same chemical nature, namely carbon. It can even be intuitively understood that the difference between the binding energies for different intermediates would likely be constant for various catalytic surfaces in this case.

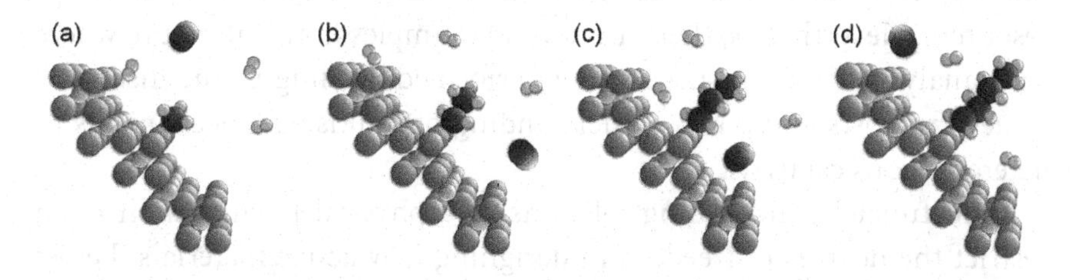

FIGURE 2.5 The origin of scaling relations. This example shows the catalytic growth of a hydrocarbon molecule at one active site. The growing chain (stages from A to D) is always adsorbed through the same kind of atom (carbon) independent from the resulting chain size.

Multiple chemical reactions have more complex mechanisms, which involve intermediate species of different nature. One classic example is the catalytic oxidation of CO molecules by molecular oxygen to form carbon dioxide, CO_2. It proceeds through the following simplified steps and intermediates (Langmuir-Hinshelwood mechanism) [32]:

$$O_{2(gas)} + 2^* \rightarrow 2O^*_{(adsorbed)}$$

$$CO_{(gas)} + {}^* \rightarrow CO^*_{(adsorbed)}$$

$$O^*_{(adsorbed)} + CO^*_{(adsorbed)} \rightarrow 2^* + CO_{2(gas)}$$

The mechanism presented above assumes the initial adsorption of the O_2 and CO gaseous species, with the dissociation of the oxygen molecule resulting in the adsorbed O*. Then O* and *CO adsorbates diffuse over the catalytic surface and meet each other to form gaseous carbon dioxide. Adsorption sites on the catalyst surface are designated as "*". It is interesting to note that two reaction intermediates are adsorbed through atoms of different chemical nature. The CO molecules are normally adsorbed through the carbon atoms, while the oxygen species form the bonds to the active centers directly. Hence, the scaling relations between two different reaction intermediates are not necessarily expected. Therefore, more than one descriptor is needed to describe the activity trends more accurately: ΔE_{CO^*} and ΔE_{O^*}. A schematic of a volcano plot for the Langmuir-Hinshelwood mechanism is shown in Figure 2.6: at least two descriptors are necessary when there is no evidence for the existence of the scaling relations between two intermediates. This certainly complicates the theoretical analysis. When more descriptors are involved, the DFT calculations typically become too "expensive" in terms of time and computational resources. Nevertheless, there are several examples in the literature where such analysis has been successfully performed, leading to the discovery of new catalysts and a new understanding of multistage mechanisms in heterogeneous catalysis.

Unfortunately, the scaling relations as a physical phenomenon often restrict the degrees of freedom in designing new active materials. Let us consider a multistage reaction of CO methanation over nickel and rhenium metallic surfaces to form methane fuel, CH_4:

$$CO + 3H_2 \rightarrow CH_4 + H_2O$$

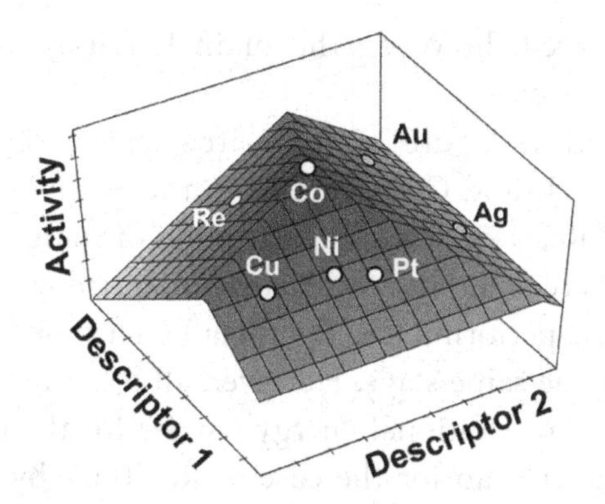

FIGURE 2.6 An example of a complex volcano plot with two descriptors (e.g., binding energies for two intermediates). One example of a catalytic reaction, where two descriptors are necessary, is the oxidation of CO molecules by oxygen, O_2, to form CO_2. In the latter case, binding energies for O* and CO* species are used as the descriptors (see text for details).

In simple terms, this reaction proceeds through several C-containing intermediates adsorbed through the carbon atoms. Hence, one can expect the influence of the scaling relations. The calculated simplified 2D energy diagram for the methanation processes is shown in Figure 2.7 (of course, H* and H_2O also contribute, and a multidimensional situation

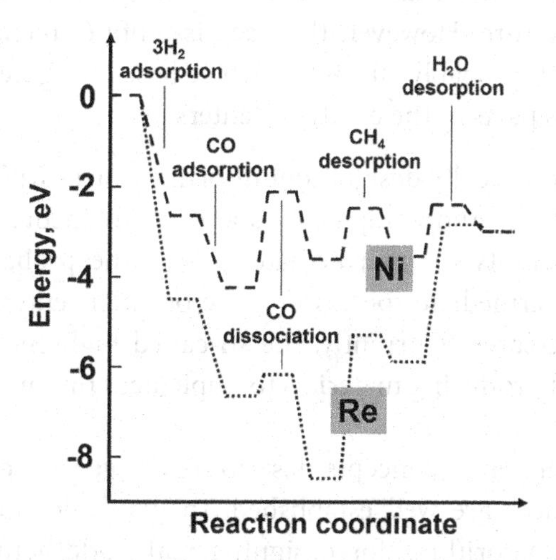

FIGURE 2.7 The consequence of the scaling relations for the CO methanation reaction over Ni and Re catalytic surfaces. It is impossible to change the binding energies of intermediates independently by changing the chemical nature of active sites. All the energy barriers are shifted simultaneously in the same direction if one replaces Re by Ni. (Adapted from [33].)

should be considered; however, the main barriers, in this case, are formed by the C-species).

As one can see from Figure 2.7, there are some energy barriers on the way from gaseous H_2 and CO reactants to the CH_4 and H_2O products. For the surface of Re, there is a deep minimum in energy, suggesting that the material binds reaction intermediates much stronger than necessary for a good catalytic performance. The metallic Ni shows a more optimal energy profile for the same stage. However, all other barriers are shifted upwards, creating an additional energy barrier for the CO dissociation step, which is higher than for the case of Re. Thus, by optimizing one stage of the multistage reaction, one can impede another step due to the scaling relations.

Taking the considerations mentioned above into account, it is essential to "overcome" the scaling relations in order to design optimal catalysts for multistage catalytic processes. There are currently no well-accepted strategies to do so. However, the following promising concepts can still be highlighted.

1. The catalytic centers should change their geometry (structure) during the reaction to provide the optimum binding for all the intermediates at each step. This approach would try to mimic some processes found in nature. However, this idea is quite challenging to implement; most probably, flexible carbon-based fragments should be included as parts of the catalytic centers.

2. One can specifically design nonuniform surfaces with different surface structures and compositions and simultaneously increase the surface mobility of intermediates. Thus, the probability that the moving intermediate species find the optimal centers for each step can be increased. Currently, the so-called high entropy alloys are considered promising materials to implement this approach [34,35].

While the fundamental concepts describing the general catalytic trends for simple surfaces are well established, they still do not directly give straightforward algorithms for designing real-world active catalysts. It should be noted here that the materials used in heterogeneous catalysis are usually nanostructured ones to maximize the available surface area for the reaction and hence increase the reaction rate (amount of products per unit of time). The catalyst nanoparticles are often immobilized on some high surface area inert or functional supports [36,37], which can also contribute to the activity by forming the support/particle boundaries [38–40]. The nanoparticles have various atoms with different coordinations and compositions, which influence the electronic structure of the catalytic sites and, hence, their adsorption properties toward the reaction intermediates. Finally, one should not forget about significant particle size effects. Figure 2.8 schematically illustrates the complexity of new catalysts' understanding and rational design, not considering that the catalysts under reaction conditions are also very dynamic systems [41].

Even such an oversimplified picture found in Figure 2.8 can explain why it is so difficult to understand how the particular catalytic system works. How to find out what are the most active sites in those catalysts [42,43]? What is the role of surface composition in various cases? What is the role of the surface structure to catalyze specific reactions? What is the contribution of surface defects? Are the catalyst-support interactions important? Real-world heterogeneous catalysts are complex systems where the resulting performance depends on many parameters. Nevertheless, in many cases, even a relatively simple theoretical analysis can significantly help in designing new heterogeneous catalysts.

Figure 2.9 shows one of the first examples of the so-called rational identification of active alloys toward methanation reaction and compares the

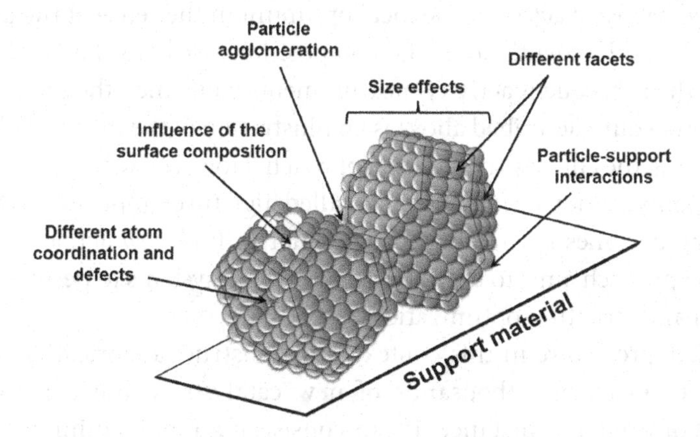

FIGURE 2.8 A schematic representation of a nanostructured supported catalyst. Even this oversimplified picture illustrates the complexity of the real-world heterogeneous catalysts, where the resulting activity can originate from the sites located at different facets, from the centers with different coordination and compositions, from the boundaries between the catalyst and support, and where the size effects can contribute significantly to the overall material performance.

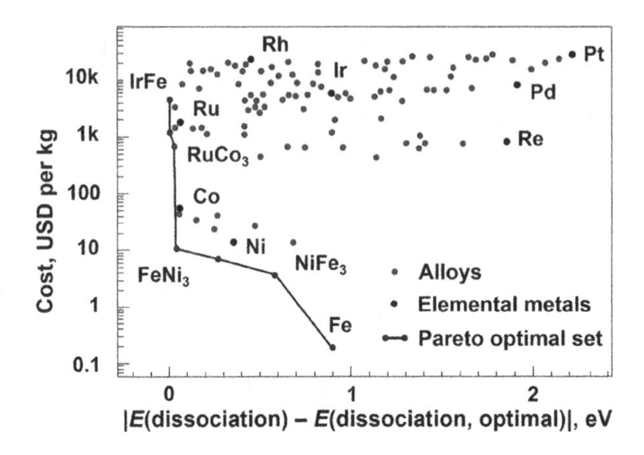

FIGURE 2.9 A price versus catalytic performance plot for the catalytic methanation reaction over a range of elemental metals and alloys. The local optimum in the activity vs. price is located close to the $FeNi_3$ material. (Adapted from [44].)

price and catalytic performance for those alloys. The DFT calculations were performed for a relatively large number of elemental metals and binary alloys. The activity maximum should be at the points where the CO dissociation barrier is the lowest (see Figure 2.7). In other words, the closer the descriptor E_{diss} (the CO dissociation energy) is to the optimum value (the smaller the value of $|E_{diss} - E_{diss, optimal}|$ in Figure 2.9 is), the better is the predicted catalytic activity. The local optimum in the sense of the activity and price can be identified at the compositions close to the $FeNi_3$ alloy. Indeed, the subsequent activity measurements confirmed the prediction.

The procedure described above is an illustration of the material identification using the entirely theoretical approach. However, with the measurement automatization progress, the so-called high throughput experimental screening becomes more and more popular to find new active catalysts. Such an approach aims to use robotic/automated systems to perform composition and structure optimizations.

The key procedure in this context is to construct a fabrication system capable of producing thousands of new catalysts within a reasonable amount of time. For instance, it can consist of a modified inkjet printer depositing a library of binary, ternary, and quaternary candidates. In the next step, it is essential to select robust and fast characterization techniques (e.g., spectroscopies) to reveal active combinations. Nevertheless, despite significant progress in both theoretical and experimental screening, it is still necessary to rationally identify a reasonable set of promising candidates for the subsequent fabrication and fundamental analyses. It is now hardly possible to cover all possible combinations without reliable criteria to include them initially in the high throughput search.

In the discussion above, it was mentioned that the real-world heterogeneous catalytic materials are nanostructured ones. Nanostructuring is necessary to increase the available surface area for the reactions. However, in most cases, the particle size will not only determine the surface area but also influence the performance of the active sites themselves [45]. This is often called a catalytic *size effect.* Interestingly, it can either increase the turnover frequency, i.e., the number of product species produced at a catalytic center per unit of time, or decrease it. Figure 2.10 illustrates this phenomenon. Imagine some catalytic reaction taking place at an extended single-crystalline surface of Au(111). From some relevant experiments and theoretical considerations, one knows that such a surface binds the intermediates of the regarded reaction too weakly. The Au(111) surface position in a volcano plot is schematically given in Figure 2.10 for such a situation. Let us also assume that one prepares nanoparticles of gold of different sizes (less than 10 nm) and narrow size distributions. With the decrease of the particle size, the fraction of atoms at the surface with a low coordination number gradually increases. At the same time, the electronic properties of the atoms with high coordination will be more and more affected by the increasing amount of atoms with lower coordination (nearest neighbor effect) [46,47]. These two factors (not counting other

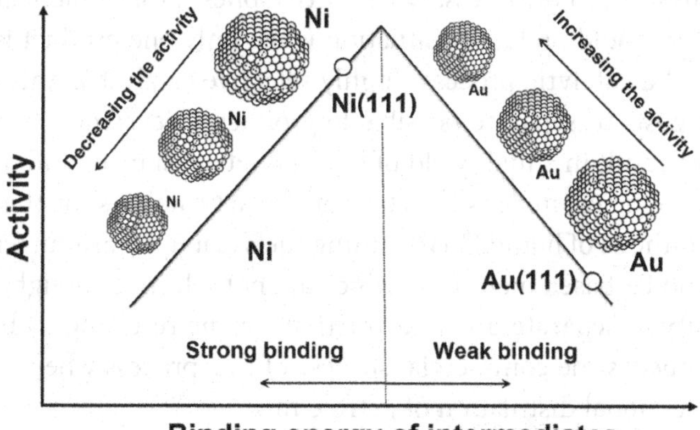

FIGURE 2.10 A schematic volcano plot illustrating the influence of the particle size on the activity for a hypothetical catalytic reaction if compared with the extended single crystal surfaces. Typically, the adsorption sites tend to bind all kinds of adsorbates stronger if the overall particle size is smaller. The effect is often noticeable when the particle size is smaller than ~10 nm.

possible contributions) will increase the surface energy of the nanoparticles and consequently increase the affinity of the surface sites toward a variety of adsorbates to minimize it. It is equivalent to the increase in the binding energy between any reaction intermediates and the catalytic centers at the surface. Considering the position of the initial Au(111) surface in the volcano plot as shown in Figure 2.10, decreasing the particle size will increase the turnover frequencies for practically all surface sites as one "moves" toward the tip of the volcano. In other words, for such kinds of reactions, where the extended surface binds reaction intermediates weaker than the optimum, the smaller the particles, the higher the specific activity is. The situation can be drastically opposite if the position of the extended surface in the corresponding volcano plot is at the stronger binding side, e.g., Ni(111) in Figure 2.10 for some reactions. When the particle size is reduced, this is equivalent to the "movement" away from the optimum. In this case, one should find a compromise between the requirement to have a high surface area of the catalyst and reasonable specific activity. Therefore, it is often not enough to extrapolate the results of the experiments and theoretical calculations obtained for the extended surfaces to the nanostructured catalytic materials.

2.3 CATALYST SELECTIVITY AND STABILITY

So far, the discussion was focused on catalyst activity only. However, very active catalysts are not necessarily selective ones. From practical considerations, one would prefer the situation when only one product is formed as a result of a catalytic process (highly selective catalysts). This excludes additional costs to separate possible byproducts. Nevertheless, an active catalyst can result in a high yield of both target substances and unwanted byproducts. One example is the state-of-the-art catalysts for the Fischer-Tropsch synthesis of liquid C-containing fuels (catalysts are currently Co-, Ni-, Ru-, and Fe-based materials), which are not selective enough. Namely, the necessity to separate different fractions of the resulting hydrocarbon mixture impedes the commercial success of this process when compared with just fractional distillation of petroleum.

DEFINITION:

A catalyst selectivity is the favoring of some specific reaction pathway between several competing pathways. A perfectly selective catalyst produces only the target product and does not produce any byproducts.

Typically, the optimization of the catalyst selectivity is a more difficult task than increasing the overall activity. For that, it is vital to assess relative changes in the energy barriers between several competing reaction pathways. A particular difficulty is the identification of key intermediates, which are primarily responsible for selectivity in the process of interest, and which property of the surface is responsible for shifting the overall reaction toward the desired product.

The development of selective catalysts is based on the elucidation of the structure-selectivity and composition-selectivity relations. The structure and composition are the main "degrees of freedom" to use in the optimization, similar to the catalytic activity. Of course, the size effects should also be taken into account. Figure 2.11a schematically shows that the same nanoparticle of a catalyst can generate different products at surface sites with different structures and coordination, as they have a different affinity to reaction intermediates.

One way to avoid such a situation is to create an abundance of facets with only one type of catalytic centers. This can be done by controlling the shape of the nanoparticles. For instance, the selectivity of silver nanoparticles toward ethylene oxidation strongly depends on the surface structure (Figure 2.11b) [48]. If the shape of the Ag species is selected to expose the Ag(111) facets maximally, ethylene is oxidized to CO_2 and H_2O, which would be desirable if ethylene is used as a fuel. If the active surface consists of only Ag(100), the main product of ethylene oxidation is ethylene oxide. This is how the same metal can be used to generate entirely different chemicals as products. Highly selective catalysts are necessary not only to produce fuels but also in catalytic converters of the cars equipped with the

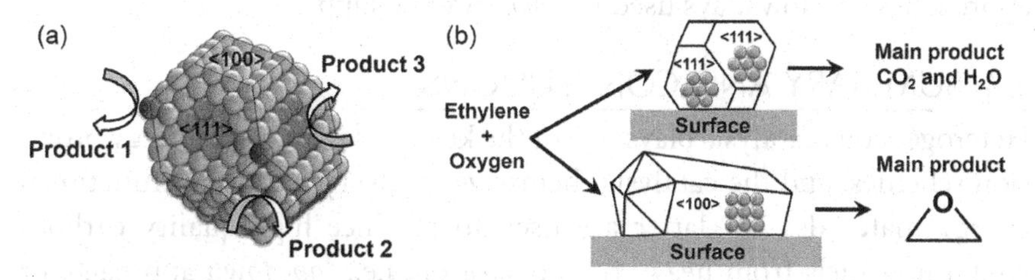

FIGURE 2.11 The importance of structure-selectivity relations in heterogeneous catalysis. (a) Schematics showing that the same catalyst nanoparticle can produce various products at the surface sites with different structures and coordination. (b) Silver nanoparticles as a particular example related to the selectivity of ethylene oxidation. (Part (b) is adapted from [48].)

standard combustion engines, which use fossil fuels. Currently, expensive palladium and other Pt-group metals are used in these converters (2–15g per car) to clean the exhaust gases from CO and other dangerous compounds and release predominantly CO_2 and H_2O into the atmosphere.

Perhaps, predicting another critical parameter characterizing the catalyst performance, namely the catalyst stability under reaction conditions, is an even more difficult task than predicting its selectivity. Stability is referred to as the ability of the catalyst to maintain the initial high activity and selectivity over time. There are, unfortunately, no well-accepted descriptors of stability, and it is typically assessed experimentally. Still, one possible parameter, which can be straightforwardly used for the initial assessment, is the heat of formation of the compounds (standard enthalpy of formation), ΔH_f, which are used as the catalytic materials, e.g., the heat of formation of alloys for the metal-based surfaces [49]. The more negative values ΔH_f for the catalysts often indicate their higher stability under similar conditions.

Finally, to mention in this chapter is the role of the catalyst-support interactions, especially for the materials used in the industrial processes. These interactions originate from the cooperative action of the different active centers that are in close vicinity to the support and the catalyst itself. For example, in the Fischer-Tropsch synthesis, introduction and control of mesoporosity in the zeolite support (to increase the number of catalyst atoms located close to the support surface) can increase the yield toward C_5–C_{11} hydrocarbons, as well as the catalyst activity and stability [50]. However, theoretical analysis of the catalyst-support interactions is, in most cases, very complex, and typically empirical trial-and-error approaches are nowadays used to select proper support.

2.4 SUMMARY AND CONCLUSIONS

Heterogeneous catalysis plays one of the key roles in current energy provision schemes, and the catalysts themselves are very important functional energy materials. The latter are used to produce high-quality carbon-containing fuels from heavy fractions of oil, i.e., *top-down* approach, or from carbon or methane, i.e., *bottom-up* processes. They are also used in the catalytic converters to clean the exhausts of cars equipped with combustion engines. While the main concepts helping at understanding the catalytic performance of numerous materials are well-known and have a long and successful history, the design of industrially relevant catalysts is complicated by the difficulties in identification of active catalytic centers, key reaction intermediates, the size effects, composition-activity relations, and catalyst-support interactions as well as issues related to the selectivity and stability of the materials under reaction conditions. With further development of theoretical and experimental high throughput screening, one can envisage a much faster identification of new catalysts for energy applications.

2.5 QUESTIONS

1. What are heterogeneous catalysis and catalysts?
2. What is the Sabatier principle? Why are the *volcano plots* widely used in heterogeneous catalysis, and what is their meaning?
3. Explain the concept of active sites.
4. Explain the reaction coordinate.
5. Look at the Figure below. The catalyst has lowered the energy barrier for a reaction, according to this scheme. Take another look at this diagram: what else did the catalyst change in this particular case?

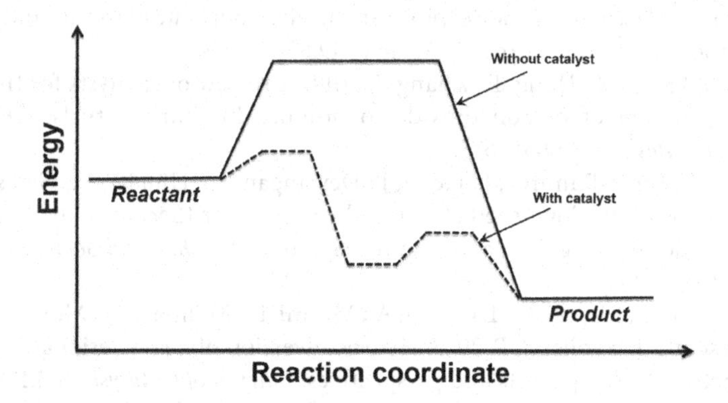

6. What are the scaling relations in heterogeneous catalysis, and what is the possible physical origin of this phenomenon?
7. What are general concepts allowing to design active, selective, and stable heterogeneous catalysts?
8. What is the turnover frequency in heterogeneous catalysis?
9. What is catalyst selectivity, and what are the general approaches to optimize it?

10. What is a possible descriptor of the catalyst stability in the case of metal alloys?
11. How does the reduction of particle size correlate with catalyst performance?

REFERENCES

1. Goeppert, A.; Czaun, M.; Jones, J.-P.; Prakash, G.K.S.; Olah, G.A. 2014. Recycling of carbon dioxide to methanol and derived products – closing the loop. *Chemical Society Reviews* 43:7995–8048.
2. Thomas, J.M. 2014. Reflections on the topic of solar fuels. *Energy & Environmental Science* 7:19–20.
3. Schlögl, R. 2016. Sustainable energy systems: The strategic role of chemical energy conversion. *Topics in Catalysis* 59:772–786.
4. Vasireddy, S.; Morreale, B.; Cugini, A.; Song, C.; Spivey, J.J. 2011. Clean liquid fuels from direct coal liquefaction: Chemistry, catalysis, technological status and challenges. *Energy & Environmental Science* 4:311–345.
5. Traa, Y. 2010. Is a renaissance of coal imminent?—challenges for catalysis. *Chemical Communications* 46:2175–2187.
6. Gao, W.; Liang, S.; Wang, R.; Jiang, Q.; Zhang, Y.; Zheng, Q.; Xie, B.; Toe, C.Y.; Zhu, X.; Wang, J.; Huang, L.; Gao, Y.; Wang, Z.; Jo, C.; Wang, Q.; Wang, L.; Liu, Y.; Louis, B.; Scott, J.; Roger, A.C.; Amal, R.; Heh, H.; Park, S.E. 2020. Industrial carbon dioxide capture and utilization: State of the art and future challenges. *Chemical Society Reviews* 49:8584–8686.
7. Qiao, J.; Liu, Y.; Hong, F.; Zhang, J. 2014. A review of catalysts for the electroreduction of carbon dioxide to produce low-carbon fuels. *Chemical Society Reviews* 43:631–675.
8. Das, S.; Pérez-Ramírez, J.; Gong, J.; Dewangan, N.; Hidajat, K.; Gates, B.C.; Kawi, S. 2020. Core–shell structured catalysts for thermocatalytic, photocatalytic, and electrocatalytic conversion of CO_2. *Chemical Society Reviews* 49:2937–3004.
9. Bellussi, G.; Rispoli, G.; Landoni, A.; Millini, R.; Molinari, D.; Montanari, E.; Moscotti, D.; Pollesel, P. 2013. Hydroconversion of heavy residues in slurry reactors: Developments and perspectives. *Journal of Catalysis* 308:189–200.
10. Prajapati, R.; Kohli, K.; Maity, S.K. 2021. Slurry phase hydrocracking of heavy oil and residue to produce lighter fuels: An experimental review. *Fuel* 288:119686.
11. Bellussi, G.; Rispoli, G.; Molinari, D.; Landoni, A.; Pollesel, P.; Panariti, N.; Millini, R.; Montanari, E. 2013. The role of MoS_2 nano-slabs in the protection of solid cracking catalysts for the total conversion of heavy oils to good quality distillates. *Catalysis Science & Technology* 3:176–182.
12. Kim, S.H.; Kim, K.D.; Lee, Y.K. 2017. Effects of dispersed MoS_2 catalysts and reaction conditions on slurry phase hydrocracking of vacuum residue. *Journal of Catalysis* 347:127–137.

13. Kim, S.H.; Kim, K.D.; Lee, Y.K. 2017. Structure and activity of dispersed Co, Ni, or Mo sulfides for slurry phase hydrocracking of vacuum residue. *Journal of Catalysis* 364:131–140.
14. Huber, G.W.; Iborra, S.; Corma, A. 2006. Synthesis of transportation fuels from biomass: Chemistry, catalysts, and engineering. *Chemical Reviews* 97:2373–2420.
15. Huber, G.W.; Corma, A. 2007. Synergies between bio-and oil refineries for the production of fuels from biomass. *Angewandte Chemie International Edition* 46:7184–7201.
16. Climent, M.J.; Corma, A.; Iborra, S. 2014. Conversion of biomass platform molecules into fuel additives and liquid hydrocarbon fuels. *Green Chemistry* 16:516–547.
17. Chorkendorff, I.; Niemantsverdriet, J.W. 2017. *Concepts of Modern Catalysis and Kinetics*. Wiley-VCH Verlag GmbH & Co. KGaA: Weinheim.
18. Thomas, J.M. 2014. Designing catalysts for tomorrow's environmentally benign processes. *Topics in Catalysis* 57:1115–1123.
19. Ostwald, W. 1902. Über Katalyse. *Physikalische Zeitschrift* 3:313–22.
20. Schlögl, R. 2015. Heterogeneous catalysis. *Angewandte Chemie International Edition* 54:3465–3520.
21. Van Santen, R.A.; Neurock, M. 2009. *Molecular Heterogeneous Catalysis: A Conceptual and Computational Approach*. John Wiley & Sons: New York.
22. Ertl, G. 2008. Reactions at surfaces: From atoms to complexity (Nobel lecture). *Angewandte Chemie International Edition* 47:3524–3535.
23. Sabatier, P. 1911. Hydrogenations et deshydrogenations par catalyse. *Berichte der Deutschen Chemischen Gesellschaft* 44:1984–2001.
24. Langmuir, I. 1922. Chemical reactions on surfaces. *Transactions of the Faraday Society* 17:607–20.
25. Taylor, H.S. 1925. A Theory of the catalytic surface. *Proceedings of the Royal Society of London Series A* 108:105–111.
26. Balandin, A.A. 1969. Modern state of the multiplet theory of heterogeneous catalysis. *Advances in Catalysis* 19:1–210.
27. Bligaard, T.; Nørskov, J.K.; Dahl, S.; Matthiesen, J.; Christensen, C.H.; Sehested, J. 2004. The Brønsted–Evans–Polanyi relation and the volcano curve in heterogeneous catalysis. *Journal of Catalysis* 224:206–217.
28. Nørskov, J.K.; Bligaard, T.; Logadottir, A.; Bahn, S.; Hansen, L.B.; Bollinger, M.; Bengaard, H.; Hammer, B., Sljivancanin, Z.; Mavrikakis, M.; Xu, Y.; Dahl, S.; Jacobsen, C.J.H. 2002. Universality in heterogeneous catalysis. *Journal of Catalysis* 209:275–278.
29. Nørskov, J.K.; Abild-Pedersen, F.; Studt, F.; Bligaard, T. 2011. Density functional theory in surface chemistry and catalysis. *Proceedings of the National Academy of Sciences* 108:937–943.
30. Abild-Pedersen, F.; Greeley, J.; Studt, F.; Rossmeisl, J.; Munter, T.R.; Moses, P.G.; Skùlason, E.; Bligaard, T.; Nørskov, J.K. 2007. Scaling properties of adsorption energies for hydrogen-containing molecules on transition-metal surfaces. *Physical Review Letters* 99:4–7.

31. Chen, Y.; Wei, J.; Duyar, M.S.; Ordomsky, V.V.; Khodakov, A.Y.; Liu, J. 2021. Carbon-based catalysts for Fischer–Tropsch synthesis. *Chemical Society Reviews* 50:2337–2366.
32. Wintterlin, J.; Völkening, S.; Janssens, T.V.W.; Zambelli, T.; Ertl, G. 1997. Atomic and macroscopic reaction rates of a surface-catalyzed reaction. *Science* 278:1931–1934.
33. Nørskov, J.; Bligaard, T.; Rossmeisl, J.; Christensen, C.H. 2009. Towards the computational design of solid catalysts. *Nature Chemistry* 1:37–46.
34. Batchelor, T.A.A.; Löffler, T.; Xiao, B.; Krysiak, O.A.; Strotkötter, V.; Pedersen, J.K.; Clausen, C.M.; Savan, A.; Li, Y.; Schuhmann, W.; Rossmeisl, J.; Ludwig, A. 2021. Complex solid solution electrocatalyst discovery by computational prediction and high-throughput experimentation. *Angewandte Chemie International Edition* 60:6932–6937.
35. Pedersen, J.K.; Batchelor, T.A.A.; Bagger, A.; Rossmeisl, J. 2020. High-entropy alloys as catalysts for the CO_2 and CO reduction reactions. *ACS Catalysis* 10: 2169–2176.
36. Munnik, P.; de Jongh, P.E.; de Jong, K.P. 2015. Recent developments in the synthesis of supported catalysts. *Chemical Reviews* 115:6687–6718.
37. Zhang, L.; Zhou, M.; Wang, A.; Zhang, T. 2020. Selective hydrogenation over supported metal catalysts: From nanoparticles to single atoms. *Chemical Reviews* 120:683–733.
38. Sankar, M.; He, Q.; Engel, R.V.; Sainna, M.A.; Logsdail, A.J.; Roldan, A.; Willock, D.J.; Agarwal, N.; Kiely, C.J.; Hutchings, G.J. 2020. Role of the support in gold-containing nanoparticles as heterogeneous catalysts. *Chemical Reviews* 120:3890–3938.
39. Lang, R.; Du, X.; Huang, Y.; Jiang, X.; Zhang, Q.; Guo, Y.; Liu, K.; Qiao, B.; Wang, A.; Zhang, T. 2020. Single-atom catalysts based on the metal–oxide interaction. *Chemical Reviews* 120:11986–12043.
40. Qin, R.; Liu, K.; Wu, Q.; Zheng, N. 2020. Surface coordination chemistry of atomically dispersed metal catalysts. *Chemical Reviews* 120:11810–11899.
41. Kalz, K.F.; Kraehnert, R.; Dvoyashkin, M.; Dittmeyer, R.; Gläser, R.; Krewer, U.; Reuter, K.; Grunwaldt, J.D. 2017. Future challenges in heterogeneous catalysis: Understanding catalysts under dynamic reaction conditions. *ChemCatChem* 9:17–29.
42. Zambelli, T.; Wintterlin, J.; Trost, J.; Ertl, G. 1996. Identification of the "active sites" of a surface-catalyzed reaction. *Science* 273:1688–1690.
43. Pfisterer, J.H.K.; Liang, Y.; Schneider, O.; Bandarenka, A.S. 2017. Direct instrumental identification of catalytically active surface sites. *Nature* 549:74–77.
44. Andersson, M.P.; Bligaard, T.; Kustov, A.; Larsen, K.E.; Greeley, J.; Johannessen, T.; Christensen, C.H.; Nørskov, J.K. 2006. Toward computational screening in heterogeneous catalysis: Pareto-optimal methanation catalysts. *Journal of Catalysis* 239:501–506.
45. Liu, L.; Corma, A. 2018. Metal catalysts for heterogeneous catalysis: From single atoms to nanoclusters and nanoparticles. *Chemical Reviews* 118:4981–5079.

46. Calle-Vallejo, F.; Tymoczko, J.; Colic, V.; Vu, Q.H.; Pohl, M.D.; Morgenstern, K.; Loffreda, D.; Sautet, P.; Schuhmann, W.; Bandarenka, A.S. 2015. Finding optimal surface sites on heterogeneous catalysts by counting nearest neighbors. *Science* 350(6257):185–189.
47. Calle-Vallejo, F.; Bandarenka. A.S. 2018. Enabling generalized coordination numbers to describe strain effects. *ChemSusChem* 11:1824–1828.
48. Christopher, P.; Linic, S. 2008. Engineering selectivity in heterogeneous catalysis: Ag nanowires as selective ethylene epoxiation catalysts. *Journal of the American Chemical Society* 130:11264–11265.
49. Greeley, J.; Stephens, I.E.L.; Bondarenko, A.S.; Johansson, T.P.; Hansen, H.A.; Jaramillo, T.F.; Rossmeisl, J.; Chorkendorff, I.; Nørskov, J.K. 2009. Alloys of platinum and early transition metals as oxygen reduction electrocatalysts. *Nature Chemistry* 1:552–556.
50. Sartipi, S.; Parashar, K.; Makkee, M.; Gascon, J.; Kapteijn, F. 2013. Breaking the Fischer–Tropsch synthesis selectivity: Direct conversion of syngas to gasoline over hierarchical Co/H-ZSM-5 catalysts. *Catalysis Science & Technology* 3:572–575.

CHAPTER 3

Electrocatalysts for Energy Provision

3.1 ELECTROCATALYSIS

The idea to use electrical current to initiate and control chemical reactions, which involve electron transfer between reacting species, the so-called *redox reactions*, has been very attractive since humankind developed stable electrical power sources for the first time. Indeed, why should one often spend costly chemicals on reducing or oxidizing other species if it is in theory possible to apply an electrical bias to do the same under much more "simplified" conditions? In other words, why not to replace common schemes (Figure 3.1a) that are based on direct electron transfer from species to species and require a very careful selection of the reaction conditions to facilitate spontaneous electron transfer. Instead, electrons can be drawn from an electron conductor polarized enough to start the redox reactions. The critical peculiarity in such a situation would be separating the reduction and oxidation processes in space, as it is schematically shown in Figure 3.1b. For such a simplified experiment, it is necessary to have at least two connected pieces of electron conductors called *electrodes* and ionically conducting media called *electrolytes*.

Interestingly, water splitting, which results in gaseous hydrogen fuel and oxygen, was one of the first reactions to implement this kind of idea. The first water *electrolyzers* were built between the 18th and 19th centuries by different scientists and in different countries. Remarkably, in multiple experiments, researchers noticed that the electrical bias, which was

DOI: 10.1201/9781003025498-3

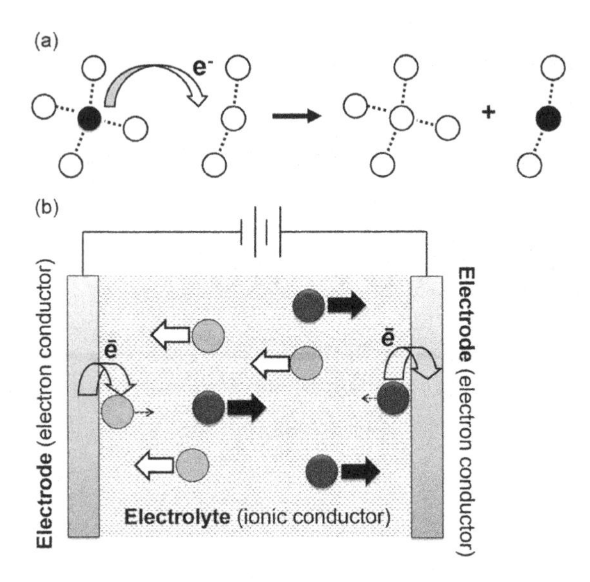

FIGURE 3.1 (a) A schematics describing a classical redox reaction, where the electron is directly transferred from one species to another to form reaction products. (b) A simplified drawing explaining the basics of electrochemical systems: there is a spatial separation of the oxidation and the reduction events. A typical electrochemical system consists of electron conductors called electrodes in contact with ion conductors called electrolytes. Both electrodes are connected via an outer circuit.

necessary to apply to start the water-splitting process, depended on the nature of the electrode materials. For instance, mercury was not a particularly good material to initiate the hydrogen evolution reaction (HER), while platinum metal was among the best metals to perform so. On the other hand, some electronically conducting oxides, like iridium oxides, were much better as anodes to produce oxygen compared to, e.g., oxides of platinum (see schematics in Figure 3.2). Later, numerous reports demonstrated that some reactions initiated by the external electrical fields do not show such a material-dependent behavior. Those are now called *outer-sphere reactions.* In these reactions, no bonds are broken or formed at the surface of electrodes, and the reactants are not specifically adsorbed. In this situation, the electrodes act just as a source or sink of electrons. There were indications that there must be both *outer-sphere* and *electrocatalytic* reactions. In the latter case, reaction kinetics largely depends on the nature of the electrode materials and the electrode surface structure, and it is also dependent on the adsorbed intermediates.

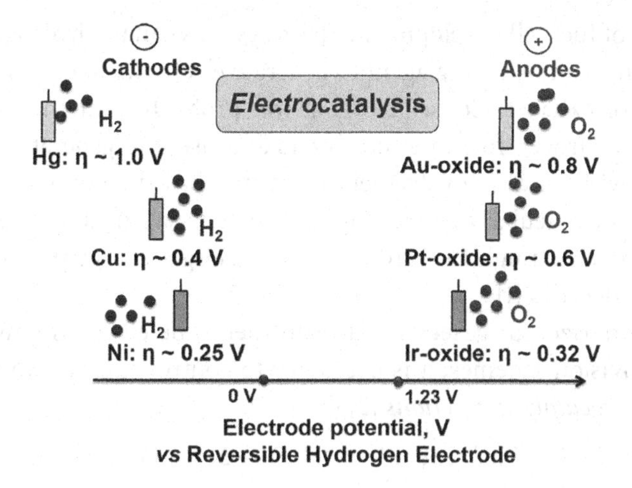

FIGURE 3.2 Water electrolysis: the HER and OER occur at different overpotentials (η, see text for details) depending on the electrode materials.

From the historical point of view, it is essential to mention that a bit later, scientists also noticed that the water-splitting reaction could be reversed directly after the electrolysis. This reverse reaction was pronounced when some electrode materials like platinum were initially used to generate hydrogen and oxygen. These facts were reported as a scientific curiosity in *ca* 1838–1842 (see Figure 3.3). That period is often considered as the

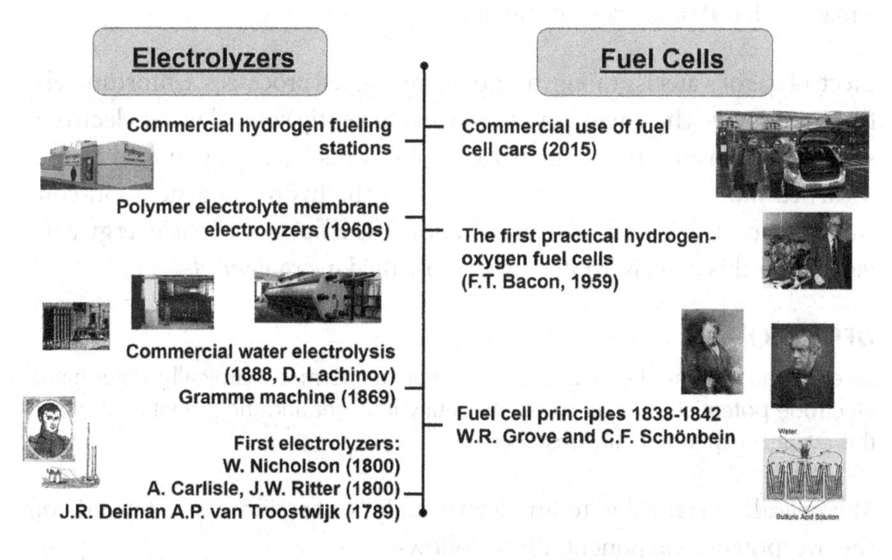

FIGURE 3.3 Some historical milestones in the development of electrolyzers and fuel cells.

beginning of fuel cell development. These systems can use hydrogen as fuel and oxygen as an oxidizing agent to generate electricity and form water as a product or exhaust. Consequently, it was probably clear that one could synthesize hydrogen fuel in electrolyzers and use it to generate electricity afterward without any involvement of carbon-based compounds in such a cycle. It is particularly interesting that the development and realization of this relatively straightforward concept at a larger energy provision scale took many decades [1].

To summarize, for gaseous hydrogen fuel to be efficiently involved in energy provision schemes, it is necessary to control at least two relatively simple *electrocatalytic reactions* [2,3]:

$$\text{Fuel generation: } 2H_2O \rightarrow 2H_2 + O_2$$

$$\text{Fuel consumption: } 2H_2 + O_2 \rightarrow 2H_2O$$

In contrast to the situation with fossil fuels, our civilization has probably enough water on the planet to generate H_2. However, in this case, the "bottlenecks" are the devices for hydrogen production, storage, and consumption [4,5].

DEFINITION:

Electrolysis of water is the decomposition of H_2O into oxygen (O_2) and hydrogen (H_2) gas induced by an electric current.

Electrolysis of water is among the most investigated processes. Unfortunately, there are severe challenges on the way to efficiently control water electrolysis and the consumption of hydrogen in fuel cells [6]. As mentioned before, it turned out that all processes involved in the hydrogen generation-consumption cycles are electrocatalytic ones. The efficiency of the energy conversion, in this case, is limited by the so-called *overpotentials*.

DEFINITION:

Overpotential, η, is the difference between a thermodynamically determined electrode potential for a redox electrocatalytic event and the potential at which this event is experimentally observed.

The overall current due to an electrocatalytic reaction often depends on the overpotential exponentially as follows:

$$i_{cat} \approx i_0 \cdot \exp\left(\left(\alpha \cdot n \cdot F \cdot |\eta|\right) / (R \cdot T)\right)$$

where i_{cat} is the current density for the catalytic reaction, i_0 is the exchange current density (when $\eta=0$), α is the so-called charge transfer coefficient or symmetry coefficient, which is often close to 0.5, n is the number of electrons involved to the electrocatalytic reaction, F is the Faraday constant, R is the universal gas constant, and T is the absolute temperature (Kelvin).

From the equation above, it is clear that even small changes in the overpotential result in a significant difference in the current. It is therefore vital to find electrode materials, which demonstrate as small overpotentials as possible. While electrolytes also play an essential role [7–13], approximately 90% of research and development in electrocatalysis are focused on designing electrodes with enhanced catalytic properties [14,15].

3.2 ELECTROCATALYSTS FOR ELECTROLYZERS

The thermodynamic analysis predicts that the minimum voltage that should be applied to an aqueous electrocatalytic system to initiate the water-splitting reaction is ~1.23 V at room temperature and atmospheric pressure. The HER should start at the cathode (see Figures 3.2 and 3.4) at 0.0 V *versus* the reversible hydrogen electrode (RHE, a special *reference*

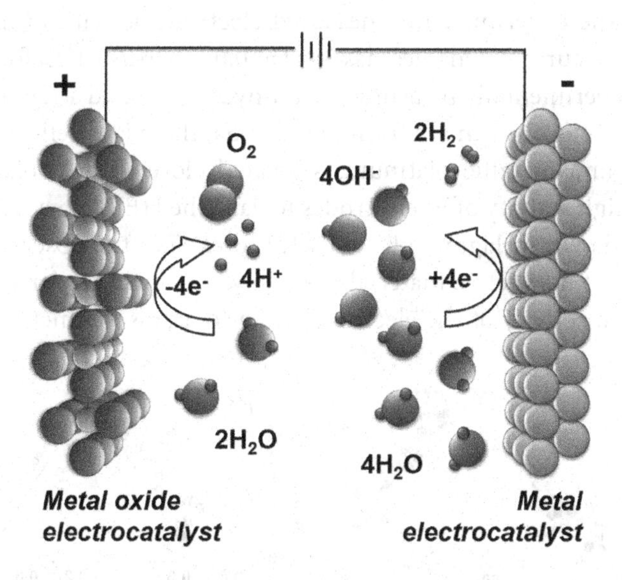

FIGURE 3.4 A scheme describing the basic principle of electrochemical water splitting. The electrode surfaces should be active to catalyze the H_2 evolution reaction at the cathode and the O_2 evolution reaction taking place at the anode. While the catalysts for hydrogen evolution are normally metals, the active anodes are electronically conducting metal oxides in most cases.

electrode and the corresponding reference scale often used in electrocatalysis). The oxygen evolution reaction (OER) should start at 1.23 V *versus* RHE. In practice, considerable overpotentials are observed for both processes; i.e., much higher than ~1.23 V external voltage has to be applied in the real-world electrolyzers. That fact leads to significant energy losses and largely increases the price of the resulting hydrogen fuel. Thus, it becomes essential to understand what determines the overpotentials.

Considering the HER, it is probably clear that the key reaction intermediates should be the hydrogen species, H*, adsorbed at the electrode surface (sign "*" designates adsorbed species). Suppose the Sabatier principle (see Chapter 2) of heterogeneous catalysis is still valid for electrocatalytic systems. In that case, the optimal catalytically active sites at the cathodes of the electrolyzers should bind those intermediates neither too strongly nor too weakly. Therefore, it should be possible to use the binding energy of the adsorbed hydrogen species as the activity descriptor and search for new materials or explain the observed activity trends based on that parameter.

Possibly, the first success of the approach mentioned above for the case of electrocatalysis was demonstrated by Sergio Trasatti in 1972 [16]. He correlated the experimentally measured electrode activities (in terms of the exchange current densities, assessed at 0.0 V *versus* RHE) for the HER with the experimentally measured metal-hydrogen bond strength (Figure 3.5a). Indeed, as one can see from Figure 3.5a, there is a well-pronounced optimum, and metallic platinum is located close to it, explaining the observed high activity of Pt electrodes toward the HER. However, the volcano plot shown in Figure 3.5a is useful to *explain* the observed trends, not to *predict* new active materials. Notably, it is much easier to measure the catalytic activity of the electrodes than to assess the metal-hydrogen

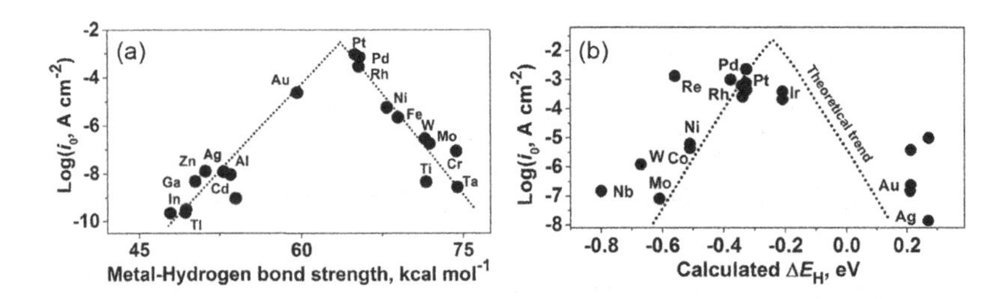

FIGURE 3.5 (a) Experimental and (b) "experimental-theoretical" volcano plots for the HER. In part (b), the binding energies of the reaction intermediates were assessed using DFT calculations. (Adapted from [16,17].)

bond strength experimentally. Primarily to address this challenge, in 2005, Jens Nørskov and his colleagues [17] introduced binding energies for the adsorbed hydrogen species, ΔE_H, calculated using density functional theory (DFT) to obtain the volcano plot for the HER (Figure 3.5b). As shown in Figure 3.5b, platinum and platinum group metals are again located close to the volcano's tip. Using DFT calculations or other quantum mechanics calculations, it is now possible to predict electrocatalytic HER activities of existing or even hypothetical surfaces. Based on such an approach, several very active platinum alloy surfaces were identified afterward [18,19].

The HER is an example of a relatively simple electrocatalytic process, where predominantly one type of reaction intermediates determines the overall electrode activity. Therefore, several catalysts, though mainly Pt-based materials [20,21], were quickly found to demonstrate very low overpotentials, which were acceptable for larger-scale commercial applications. The situation is much more complicated at the anode side of the electrolyzers, where the OER takes place. The latter has a multistage mechanism, with at least three different types of adsorbed intermediates, namely *OH, *O, and *OOH. A schematic energy diagram for the OER is shown in Figure 3.6. Several energy barriers, not only one as for the case of the HER, can be expected on the way to obtain gaseous oxygen.

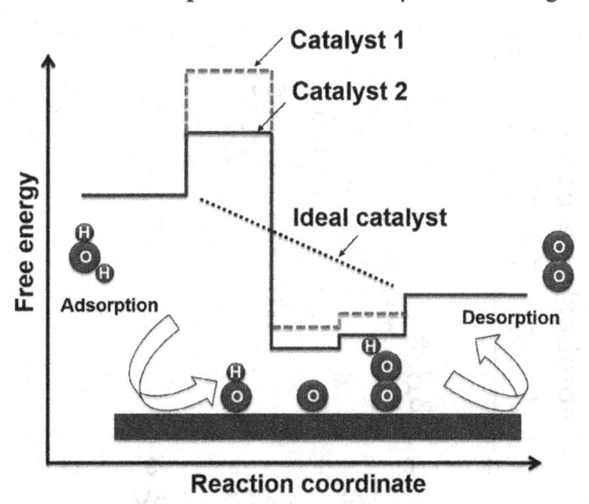

FIGURE 3.6 Scaling relations largely determine the activity of electrocatalysts in the case of multistage reactions. This scheme shows that they are also important for the multistage OER at the electrolyzers' anodes (at a constant electrode potential). It is not possible to change the energy barrier for a particular stage independently without corresponding changes in the other steps.

Interestingly, similar to the situation in classical heterogeneous catalysis, the scaling relations start to play a significant role here, largely complicating identifying the most optimal catalytic surface. As shown in Figure 3.6, it is unfortunately impossible to independently change one energy barrier without introducing corresponding changes in all the other ones by changing the electrode material if the active sites do not essentially change their geometry during the reaction steps.

Scaling relations make the situation in the case of multistage reactions very complicated. Suppose the geometry and the electronic properties of a catalytic center do not optimize themselves during the reaction. In that case, even the most active catalyst might only start to work far from the thermodynamically predicted equilibrium electrode potentials. Normally, the more stages in an electrocatalytic reaction take place, the higher the probability that the most active catalyst will be well away from the optimum predicted by classical thermodynamics [22,23]. This is demonstrated in the volcano plot for the OER shown in Figure 3.7. Ruthenium dioxide, RuO_2, is located at the tip of the theoretical plot. However, it still demonstrates an overpotential of *ca* 0.25–0.3V [24,25].

In Figure 3.7, one can see that nickel-based and iridium oxides should also be relatively active toward the OER [26–29]. Indeed, usually, those oxides are used in real-world alkaline and polymer electrolyte membrane

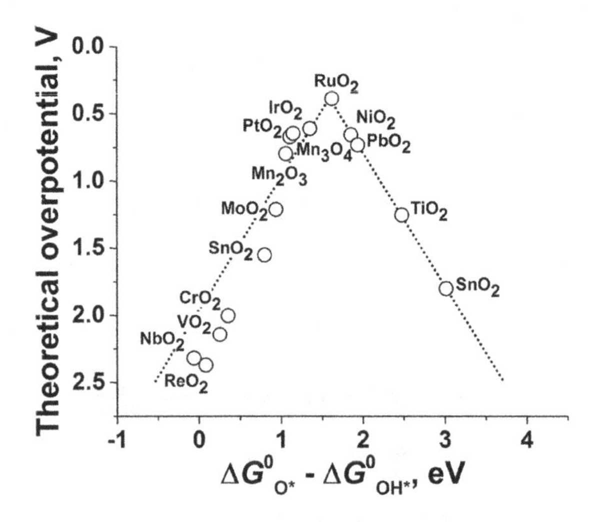

FIGURE 3.7 A theoretical volcano plot for the OER. Note that the activity descriptor in this particular case is not the binding energy for just one intermediate. It is the difference between the binding energies of two important OER intermediates (O* and OH*). (The data are from [35].)

(PEM) [30] electrolyzers, respectively. Why not RuO_2? It is simply not stable under the very harsh operational conditions of the electrolyzer anodes and often requires IrO_x or Sr^{2+} additives to increase its stability [31–34]. Here, one meets the situation when the best activity alone is not enough. Activity *versus* stability determines successful commercial applications.

The choice of electrocatalysts for electrolyzers also depends on the type of electrolytes and electrolyte composition. There are two main types of low-temperature electrolyzers: They use either KOH-containing (so-called alkaline) or acidic polymer (PEM) electrolytes [36]. Schematics of the working principles of those devices are shown in Figure 3.8.

The mechanisms of the HER and OER are slightly different for them, as the pH of the electrolytes and, therefore, the nature of reactants are dissimilar, as explained in Figure 3.8. However, the overall water splitting reaction results in the same products. Historically, alkaline electrolyzers were one of the first relatively successful commercial ones for several reasons. First of all, one can use affordable aqueous KOH solutions. From another point of view, high pH values of the electrolytes allow the implementation of cheap NiO_x-based electrocatalysts for the OER. They are stable in such media and, at the same time, are located close to the tip of the OER volcano plot (Figure 3.7), i.e., belong to the most active materials.

The first PEM-electrolyzers appeared in the 1960s. They can be operated under higher current densities and can produce more hydrogen per electrolyzer per unit of time. For the PEM-electrolyzers, stability issues

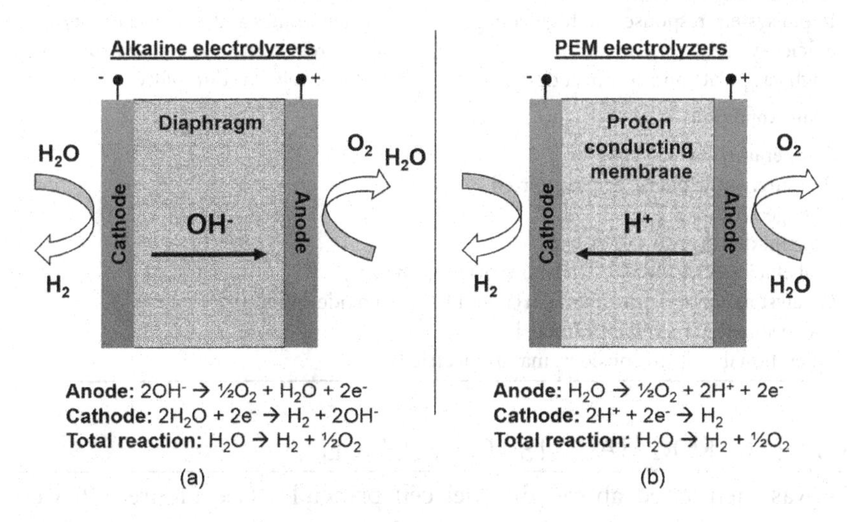

FIGURE 3.8 Operating principles of (a) alkaline and (b) PEM water electrolysis cells.

related to the nickel oxides, however, prevent their use. The harsh corrosive conditions force selection, for instance, highly stable IrO_2. Such a catalyst is also located close to the optimum in the OER volcano (Figure 3.7). Unfortunately, it is scarce and expensive.

Some advantages and disadvantages of the alkaline and PEM electrolyzers are also summarized in Tables 3.1 and 3.2.

TABLE 3.1 Conventional Alkaline Electrolyzers

Advantages	Disadvantages
• Probably, fewer development efforts are required • Successfully demonstrated over large and small scales	• Low current density and lower production rate • Lower efficiency compared to other technologies
Working conditions • Temperature: 40°C–90°C • Pressure: atmospheric or pressurized • Voltage: ~2.0 V • Current density: 0.13–0.23 A/cm^2 • Cell efficiency: 60%–80% (in laboratories up to 93%) • Catalyst material on the anode: $NiO_x/CoO_y/FeO_z$ • Catalyst material on the cathode: Ni/C • Operational costs to consider are mainly those related to electricity	

TABLE 3.2 PEM Electrolyzers

Advantages	Disadvantages
• High current densities • Rapid system response and high voltage efficiency • High gas purity and compact design	• High costs of components, such as membranes and electrocatalysts • Corrosion in an acidic environment • Relatively low durability
Working conditions • Temperature: 50°C–80°C • Pressure: atmospheric or pressurized • Voltage: ~1.8–2.2 V • Current density: up to 2 A/cm^2 • Cell efficiency: 60%–80% (in laboratories up to 90%) • Catalyst material on the anode: IrO_x and Ir-based oxide systems • Catalyst material on the cathode: Pt • Operational costs to consider: mainly electricity	

3.3 ELECTROCATALYSTS FOR FUEL CELLS

As was mentioned above, the fuel cell principles (see Figure 3.9) were demonstrated in 1838–1842. However, fuel cells were considered as just a curiosity until the early 1940s. At that time, Francis Bacon proposed the

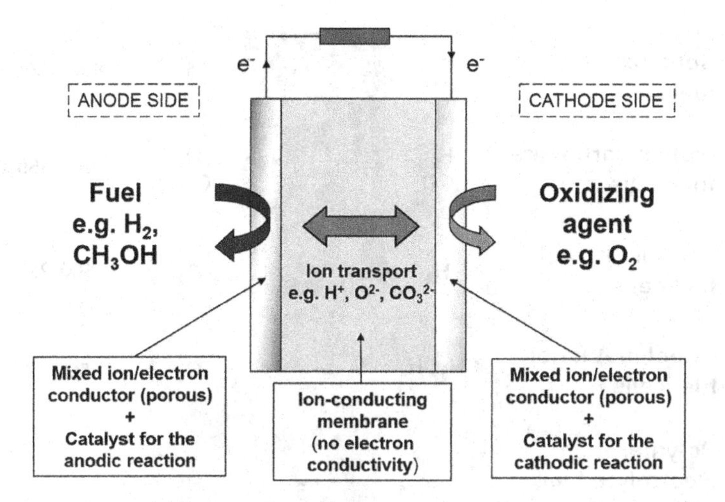

FIGURE 3.9 A scheme describing the working principles of fuel cells. The electrode surfaces should be active to catalyze spontaneous fuel oxidation reactions at the anode and the corresponding oxidant reduction reactions at the cathode.

use of fuel cells in submarines. He demonstrated a successful six-kilowatt fuel cell in 1959.

DEFINITION:

A fuel cell is a device that produces electricity using a spontaneous spatially separated *redox* reaction between a source fuel and an oxidant. The source fuel can be potentially anything that can be oxidized, including hydrogen, methane, methanol, diesel fuel, etc. The oxidant can be, e.g., atmospheric oxygen. Thus, fuel cells can produce electricity continuously as long as the fuel and oxidant are provided.

A central part of a fuel cell is an electrolyte, which can be liquid or solid (the properties and development principles of the electrolytes for energy applications will be considered in the next chapter). Two types of reactions taking place at the electrodes should be catalyzed. Fuel is oxidized at the anode side, and the electrode typically needs to be porous to provide fast transport of reactants and products. Similarly, on the cathode side, the oxidizing agent should be reduced at the catalyst surface. Almost all fuel cells use molecular oxygen at the cathode side for practical reasons, so that the relatively slow oxygen reduction reaction (ORR) has to be catalyzed.

Depending on the electrolytes or used fuels, one can distinguish several main types of fuel cells. High-temperature fuel cells include solid oxide fuel cells (SOFCs), molten carbonate, and phosphoric acid fuel cells (Figure 3.10). Out of them, SOFCs are nowadays the most promising to be widely

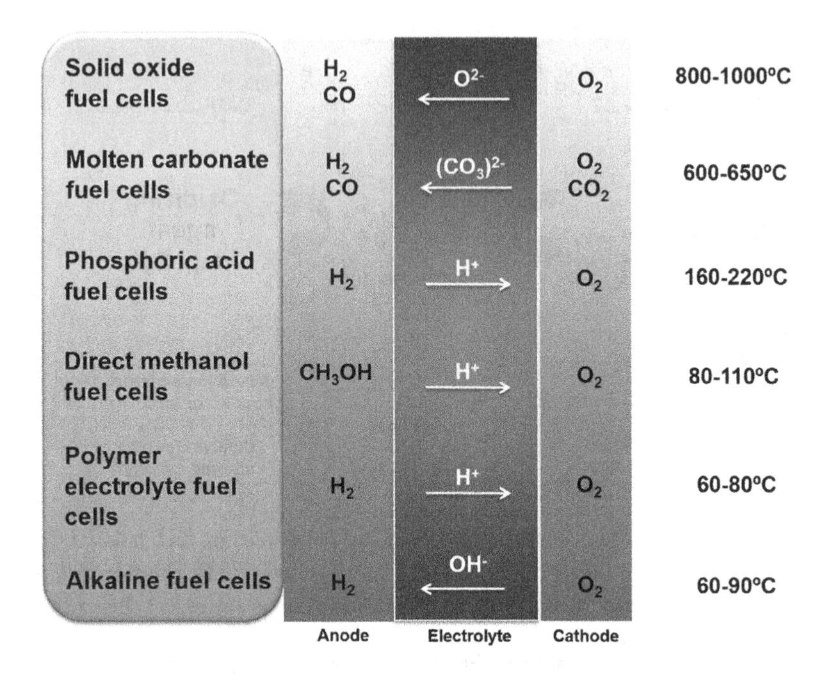

FIGURE 3.10 Common types of fuel cells, their operating temperature ranges, and the type of electrolytes used. Very often, the type of the electrolyte determines the choice of the functional energy materials and the name of the fuel cells.

commercialized. This is mainly due to two factors. The first one is that they use solid oxygen conducting membranes and no liquids, facilitating their operation. The second reason is that due to the high operating temperatures of *ca.* 800°C, almost all fuels and cheap catalysts can be involved. Typically, composites containing nickel metal and ceramics are used at the anode sides, and various perovskite materials are active enough to reduce oxygen at the cathode side. In principle, one can say that for SOFCs, the main problem is to find solid electrolytes, conductive enough (see the next chapter) at as low as possible temperatures to optimize the use of the whole system.

The most commercially promising *low-temperature* fuel cells are the PEM fuel cells (PEMFCs), direct methanol, and alkaline fuel cells. Out of them, PEMFCs operating at temperatures of *ca* 80°C are nowadays feasible for transportation, *i.e.*, in cars [37,38], buses [39] trains [40], and even airplanes [41]. However, low operating temperatures require specific catalysts to promote slow ORR at the cathodes [42]. The advantages of SOFCs and PEMFCs are summarized in Table 3.3. As the PEMFCs should potentially be the most frequently met due to a large "automotive potential", as it is also now, in the following, we focus on the catalysis of the ORR at PEMFC cathodes.

TABLE 3.3 Advantages of PEM and Solid Oxide Fuel Cells

PEM Fuel Cells	Solid Oxide Fuel Cells
• Operate at relatively low temperatures, *ca* 80°C • Have high power density • Can vary output quickly • Well-suited to the automotive applications, where a quick startup is required	• Very flexible with respect to the used fuels. They can reform methane internally, use even carbon monoxide as a fuel, and tolerate common fossil fuels' impurities. • The high operating temperatures permit the use of lower-cost catalysts

Similar to the OER taking place in the electrolyzers, the reverse reaction, the ORR, has at least three types of similar intermediates: OH*, O*, and OOH*. Unfortunately, scaling relations also limit the design of the best possible electrocatalysts for this reaction. The most active electrocatalytic surfaces were identified to be Pt and its alloys, as illustrated in Figure 3.11. However, even those materials demonstrate the overpotentials close to 0.3 V. Moreover, the most active alloy, Pt_3Ni [43–45], is not stable enough to use it in fuel cell vehicles so far.

Let us also consider a schematic energy diagram for the ORR as a function of the electrode potential in the case of Pt-based materials (Figure 3.12). As one can see from the Figure, at the thermodynamically predicted equilibrium potential of 1.23 V, the energy diagram discloses two significant barriers (designated by arrows in Figure 3.12) on the way from molecular oxygen to water, suggesting that the rate of the reaction should be negligibly low, if not taking place at all within the timeframes of conventional experiments. When the electrode potential is moved toward

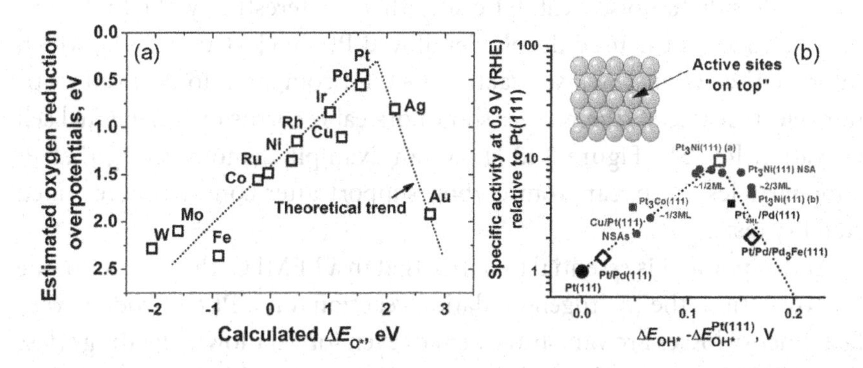

FIGURE 3.11 (a) A theoretical volcano plot for the ORR in the case of pure metals with closely arranged surface atoms, i.e., the surfaces resembling the fcc(111) surfaces. (Adapted from [46]). (b) A closer look at the tip of the ORR volcano, where Pt-alloys are used to reach the optimum. (Adapted from [47].)

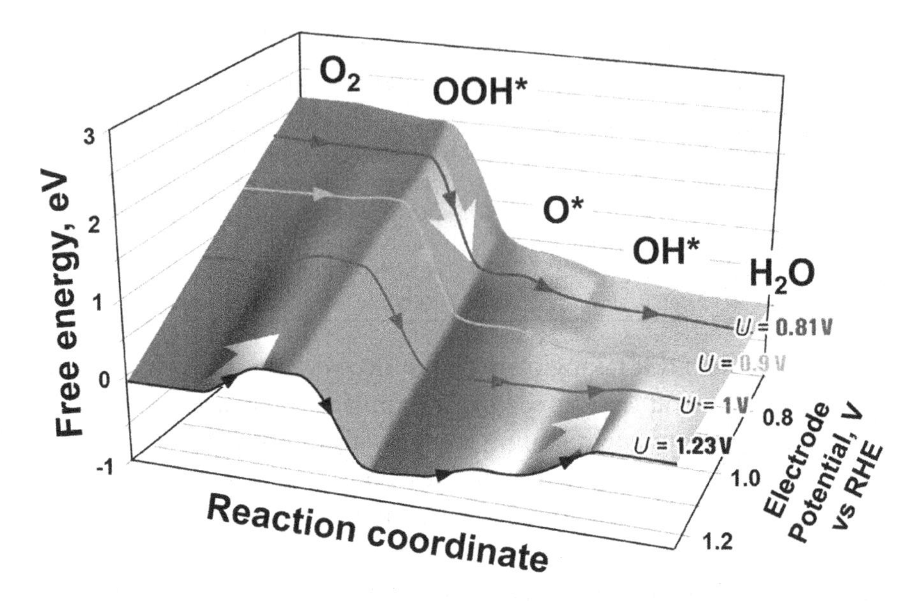

FIGURE 3.12 Transformations in the energy diagram related to the ORR with the electrode potentials at fuel cell cathodes. The lower potentials facilitate oxygen electroreduction. (Adapted from [48].)

more negative values, the barriers become smaller and almost disappear when the potential approaches *ca* 0.9 V.

Certain stability issues so far prohibit a wider use of Pt alloys with late transition metals, even though they are positioned at the top of the activity volcano plot. For instance, ions of solute elements like Ni^{2+} can be released from the degrading cathode electrocatalysts and migrate toward the anode side to poison catalytic sites there. Interestingly, the first commercial fuel cell car used deeply de-alloyed Pt/Co electrocatalysts, which improve the activity by just a factor of ~1.8 if compared to Pt. Even platinum electrocatalysts themselves show noticeable corrosion during fuel cell operation [49–51]. Figure 3.13 shows an example demonstrating that Pt nanoparticles disappear from carbon support after common accelerated stability tests.

At this point, it is essential to notice that in a PEMFC, the kinetics of the fuel oxidation, the hydrogen oxidation reaction, on a Pt electrode is very fast. Energy losses are vanishingly small even for very low Pt loadings (less than 5 mV in terms of voltage losses with Pt anode loadings of 0.05 mg cm^{-2} or, in some cases, ~0.01 mg cm^{-2}).

Finally, what are the perspectives of fuel cell applications? Many concerns in the past were related to the availability and price of the main

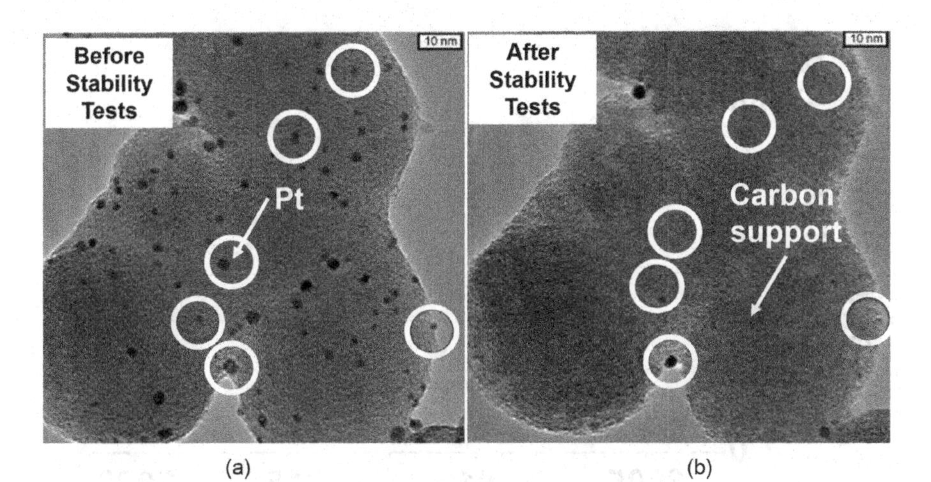

FIGURE 3.13 Stability related to platinum nanoparticles catalyzing the ORR at the fuel cell cathodes. Even noble Pt metal species can be dissolved under realistic operation conditions of the PEM fuel cells. The figure shows the catalyst (a) before its use and (b) the degraded material with only a few particles left at the carbon support surface. (Adapted from [52].)

catalyst: platinum. However, nowadays, in simple terms, one can compare how much platinum or platinum group metals is needed for one fuel cell vehicle and one catalytic converter of a typical combustion engine car. Figure 3.14 compares such values. One can see that the production of one fuel cell car for private use will most likely require not much more Pt or platinum group metals than manufacturing one catalytic converter of a "normal" combustion engine car. Therefore, one can assume a substantial increase in the further commercialization of this technology.

3.4 SUMMARY AND CONCLUSIONS

Efficient electrocatalysis of the key energy reactions is probably one of the answers to various challenges of future and current energy provision schemes. While only ~4% of hydrogen is nowadays produced by water electrolysis, there is a growing understanding that electrocatalysts will likely play a vital role in energy material science. With this respect, one needs to control at least two electrocatalytic electrode processes: hydrogen production through water electrolysis and hydrogen consumption in fuel cells. In both cases, the nonoptimal electrodes, which have adsorbed oxygen intermediates, are the impediments to implement the hydrogen economy concept. While the observed activity trends are understood, the ways

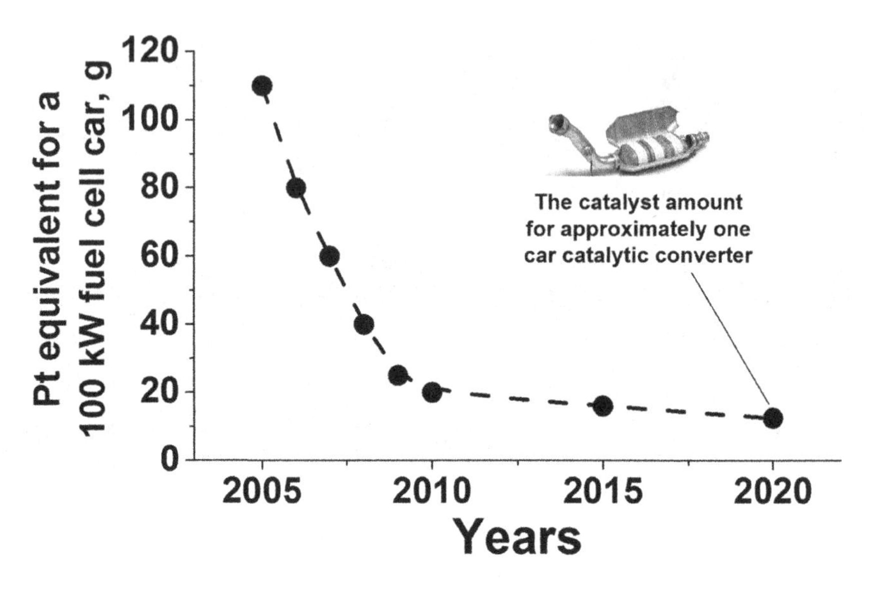

FIGURE 3.14 The average platinum group metal loading for a 100 kW fuel cell car. The amount of the platinum group metals is currently comparable with that needed for the catalytic converters of the combustion engine vehicles to follow the safety rules in a number of countries. (Adapted from [53].)

of how to design active electrocatalysts are known, and new approaches are also suggested, for instance, using high-entropy alloys [54], numerous stability issues related to the ORR and OER catalysts are to be further addressed [55,56].

3.5 QUESTIONS

1. How can the electrical current be used to control redox reactions?
2. Explain the basic principles of water electrolysis. What is an overpotential?
3. What are electrocatalysis and electrocatalysts? Analyze the differences between alkaline and acidic water electrolysis.
4. Explain the working principles and name the main components of alkaline and PEM electrolyzers.
5. Name state-of-the-art electrocatalysts for alkaline and acidic water electrolysis.
6. What are the reaction intermediates for the HER and OER?

7. Explain the concept of volcano plots for electrocatalysis, taking the H_2 and O_2 evolution reactions as an example.
8. What are fuel cells, and what are the basic principles of their operation?
9. What are the main types of fuel cells?
10. Name state-of-the-art electrocatalysts for PEM fuel cells and SOFCs.
11. What causes the main losses in PEM fuel cells?
12. Analyze the advantages and disadvantages of high-temperature fuel cells.

REFERENCES

1. Greeley, J.; Markovic, N.M. 2012. The road from animal electricity to green energy: Combining experiment and theory in electrocatalysis. *Energy & Environmental Science* 5:9246–9256.
2. Stephens, I.E.L.; Rossmeisl, J.; Chorkendorff, I. 2016. Toward sustainable fuel cells. *Science* 354:1378–1379.
3. Gasteiger, H.A.; Marković, N.M. 2009. Just a dream - or future reality? *Science* 324:48–49.
4. Stamenkovic, V.R.; Strmcnik, D.; Lopes, P.P.; Markovic, N.M. 2017. Energy and fuels from electrochemical interfaces. *Nature Materials* 16:57–69.
5. Katsounaros, I.; Cherevko, S.; Zeradjanin, A.R.; Mayrhofer, K.J.J. 2014. Oxygen electrochemistry as a cornerstone for sustainable energy conversion. *Angewandte Chemie International Edition* 53:102–121.
6. Buttler, A.; Spliethoff, H. 2018. Current status of water electrolysis for energy storage, grid balancing and sector coupling via power-to-gas and power-to-liquids: A review. *Renewable and Sustainable Energy Reviews* 82:2440–2454.
7. Lopes, P.P.; Strmcnik, D.; Jirkovsky, J.S.; Connell, J.G.; Stamenkovic, V.; Markovic, N. 2016. Double layer effects in electrocatalysis: The oxygen reduction reaction and ethanol oxidation reaction on Au(111), Pt(111) and Ir(111) in alkaline media containing Na and Li cations. *Catalysis Today* 262:41–47.
8. Strmcnik, D.; Kodama, K.; van der Vliet, D.; Greeley, J.; Stamenkovic, V.R.; Marković, N.M. 2009. The role of non-covalent interactions in electrocatalytic fuel-cell reactions on platinum. *Nature Chemistry* 1:466–472.
9. Garlyyev, B.; Xue, S.; Pohl, M.D.; Reinisch, D.; Bandarenka, A.S. 2018. Oxygen electroreduction at high-index Pt electrodes in alkaline electrolytes: A decisive role of the alkali metal cations. *ACS Omega* 3:15325–15331.
10. Xue, S.; Garlyyev, B.; Watzele, S.; Liang, Y.; Fichtner, J.; Pohl, M.D.; Bandarenka, A.S. 2018. Influence of alkali metal cations on the hydrogen

evolution reaction activity of Pt, Ir, Au and Ag electrodes in alkaline electrolytes. *ChemElectroChem* 5:2326–2329.
11. Tymoczko, J.; Colic, V.; Ganassin, A.; Schuhmann, W.; Bandarenka, A.S. 2015. Influence of the alkali metal cations on the activity of Pt(111) towards model electrocatalytic reactions in acidic sulfuric media. *Catalysis Today* 244:96–102.
12. Suntivich, J.; Perry, E.E.; Gasteiger, H.A.; Shao-Horn, Y. 2013. The influence of the cation on the oxygen reduction and evolution activities of oxide surfaces in alkaline electrolyte. *Electrocatalysis* 4:49–55.
13. Durst, J.; Siebel, A.; Simon, C.; Hasché, F.; Herranz, J.; Gasteiger, H.A. 2014. New insights into the electrochemical hydrogen oxidation and evolution reaction mechanism. *Energy & Environmental Science* 7:2255–2260.
14. Colic, V.; Pohl, M.; Scieszka, D.; Bandarenka, A.S. 2016. Influence of the electrolyte composition on the activity and selectivity of electrocatalytic centers. *Catalysis Today* 262:24–35.
15. Garlyyev, B.; Fichtner, J.; Piqué, O.; Schneider, O.; Bandarenka, A.S.; Calle-Vallejo, F. 2019. Revealing the nature of active sites in electrocatalysis. *Chemical Science* 10:8060–8075.
16. Trasatti, S. 1972. Work function, electronegativity, and electrochemical behaviour of metals: III. Electrolytic hydrogen evolution in acid solutions. *Journal of Electroanalytical Chemistry* 39:163–184.
17. Nørskov, J.K.; Bligaard, T.; Logadottir, A.; Kitchin, J.R.; Chen, J.G.; Pandelov, S.; Stimming, U. 2005. Trends in the exchange current for hydrogen evolution. *Journal of the Electrochemical Society* 152(3):J23–J26.
18. Greeley, J., Jaramillo, T., Bonde, J.; Chorkendorff, I.; Norskov, J.K. 2006. Computational high-throughput screening of electrocatalytic materials for hydrogen evolution. *Nature Materials* 5:909–913.
19. Tymoczko, J.; Calle-Vallejo, F.; Schuhmann, W.; Bandarenka, A.S. 2016. Making the hydrogen evolution reaction in polymer electrolyte membrane electrolysers even faster. *Nature Communications* 7:10990.
20. Hansen, J.N.; Prats, H.; Toudahl, K.K.; Secher, N.M.; Chan, K.; Kibsgaard, J.; Chorkendorff, I. 2021. Is there anything better than Pt for HER? *ACS Energy Letters* 6:1175–1180.
21. Strmcnik, D.; Lopes, P.P.; Genorio, B.; Stamenkovic, V.R.; Markovic, N.M. 2016. Design principles for hydrogen evolution reaction catalyst materials. *Nano Energy* 29:29–36..
22. Pittkowski, R.; Krtil, P.; Rossmeisl, J. 2018. Rationality in the new oxygen evolution catalyst development. *Current Opinion in Electrochemistry* 12:218–224.
23. Busch, M.; Halck, N.B.; Kramm, U.I.; Siahrostami, S.; Krtil, P.; Rossmeisl, J. 2016. Beyond the top of the volcano?–A unified approach to electrocatalytic oxygen reduction and oxygen evolution. *Nano Energy* 29:126–135.
24. Stoerzinger, K.A.; Diaz-Morales, O.; Kolb, M.; Rao, R.R.; Frydendal, R.; Qiao, L.; Wang, X.R.; Halck, N.B.; Rossmeisl, J.; Hansen, H.A.; Vegge, T.; Stephens, I.E.L.; Koper, M.T.M.; Shao-Horn, Y. 2017. Orientation-dependent oxygen evolution on RuO_2 without lattice exchange. *ACS Energy Letters* 2:876–881.

25. Rao, R.R.; Kolb, M.J.; Halck, N.B.; Pedersen, A.F.; Mehta, A.; You, H.; Stoerzinger, K.A.; Feng, Z.; Hansen, H.A.; Zhou, H.; Giordano, L.; Rossmeisl, J.; Vegge, T.; Chorkendorff, I.; Stephens, I.E.L.; Shao-Horn, Y. 2017. Towards identifying the active sites on RuO_2 (110) in catalyzing oxygen evolution. *Energy & Environmental Science* 10:2626–2637.
26. Subbaraman, R.; Tripkovic, D.; Chang, K.C.; Strmcnik, D.; Paulikas, A.P., Hirunsit, P.; Chan, M.; Greeley, J.; Stamenkovic, V.; Markovic, N.M. 2012. Trends in activity for the water electrolyser reactions on 3d M (Ni, Co, Fe, Mn) hydr(oxy)oxide catalysts. *Nature Materials* 11:550–557.
27. Diaz-Morales, O.; Ledezma-Yanez, I.; Koper, M.T.M.; Calle-Vallejo, F. 2015. Guidelines for the rational design of Ni-based double hydroxide electrocatalysts for the oxygen evolution reaction. *ACS Catalysis* 5:5380–5387.
28. Roy, C.; Sebok, B.; Scott, S.B.; Fiordaliso, E.M.; Sørensen, J.E.; Bodin, A.; Trimarco, D.B.; Damsgaard, C.D.; Vesborg, P.C.K.; Hansen, O.; Stephens, I.E.L.; Kibsgaard, J.; Chorkendorff, I. 2018. Impact of nanoparticle size and lattice oxygen on water oxidation on $NiFeO_xH_y$. *Nature Catalysis* 1:820–829.
29. Dionigi, F.; Strasser, P. 2016. NiFe-based (oxy)hydroxide catalysts for oxygen evolution reaction in non-acidic electrolytes. *Advanced Energy Materials* 6:1600621.
30. Carmo, M.; Fritz, D.L. Mergel, J.; Stolten, D. 2013. A comprehensive review on PEM water electrolysis. *International Journal of Hydrogen Energy* 38:4901–4934.
31. Roy, C.; Rao, R.R.; Stoerzinger, K.A.; Hwang, J.; Rossmeisl, J.; Chorkendorff, I.; Shao-Horn, Y.; Stephens, I.E.L. 2018. Trends in activity and dissolution on RuO_2 under oxygen evolution conditions: Particles versus well-defined extended surfaces. *ACS Energy Letters* 3: 2045–2051.
32. Escudero-Escribano, M.; Pedersen, A.F.; Paoli, E.A.; Frydendal, R.; Friebel, D.; Malacrida, P.; Rossmeisl, J.; Stephens, I.E.L.; Chorkendorff, I. 2018. Importance of surface IrO_x in stabilizing RuO_2 for oxygen evolution. *The Journal of Physical Chemistry B* 122:947–955.
33. Chang, S.H.; Danilovic, N.; Chang, K.C.; Subbaraman, R.; Paulikas, A.P.; Fong, D.D.; Highland, M.J.; Baldo, P.M.; Stamenkovic, V.R.; Freeland, J.W.; Eastman, J.A.; Markovic, N.M. 2014. Functional links between stability and reactivity of strontium ruthenate single crystals during oxygen evolution. *Nature Communications* 5:1–9.
34. Frydendal, R.; Paoli, E.A.; Knudsen, B.P.; Wickman, B.; Malacrida, P.; Stephens, I.E.L.; Chorkendorff, I. 2014. Benchmarking the stability of oxygen evolution reaction catalysts: The importance of monitoring mass losses. *ChemElectroChem* 1:2075–2081.
35. Man, I.C.; Su, H.Y.; Calle-Vallejo, F.; Hansen, H.A.; Martínez, J.I.; Inoglu, N.G.; Kitchin, J.; Jaramillo, T.F.; Nørskov, J.K.; Rossmeisl, J. 2011. Universality in oxygen evolution electrocatalysis on oxide surfaces. *ChemCatChem* 3:1159–1165.
36. Schalenbach, M.; Tjarks, G.; Carmo, M.; Lueke, W.; Mueller, M.; Stolten, D. 2016. Acidic or alkaline? Towards a new perspective on the efficiency of water electrolysis. *Journal of The Electrochemical Society* 163:F3197–F3208.

37. Wilberforce, T.; El-Hassan, Z.; Khatib, F.N.; Al Makky, A.; Baroutaji, A.; Carton, J.G.; Olabi, A.G. 2017. Developments of electric cars and fuel cell hydrogen electric cars. *International Journal of Hydrogen Energy* 42:25695–25734.
38. Yoshida, T.; Kojima, K. 2015. Toyota MIRAI fuel cell vehicle and progress toward a future hydrogen society. *Electrochemical Society Interface* 24:45–49.
39. Hyundai Motor's Elec City Fuel Cell Bus Begins Trial Service in Munich, Germany. https://www.hyundai.news/eu/articles/press-releases/hyundais-elec-city-fuel-cell-bus-trial-service-in-germany.html (Accessed: July 2021).
40. Siemens, DB plan fuel cell Mireo Plus H train. 2020. *Fuel Cells Bulletin* December: 6.
41. Fuel cell aircraft presented at Stuttgart Airport.Eco-friendly and silent propulsion system receives permit for test flight. https://www.uni-ulm.de/en/in/faculty-of-engineering-computer-science-and-psychology/in-detailseiten/news-detail/article/fuel-cell-aircraft-presented-at-stuttgart-airport-1/ (Accessed: July 2021).
42. Kulkarni, A.; Siahrostami, S.; Patel, A.; Nørskov, J.K. 2018. Understanding catalytic activity trends in the oxygen reduction reaction. *Chemical Reviews* 118:2302–2312.
43. Stamenkovic, V.R.; Fowler, B.; Mun, B.S.; Wang, G.; Ross, P.N.; Lucas, C.A.; Marković, N.M. 2007. Improved oxygen reduction activity on $Pt_3Ni(111)$ via increased surface site availability. *Science* 5811:493–497.
44. Chen, C.; Kang, Y.; Huo, Z.; Zhu, Z.; Huang, W.; Xin, H.L.; Snyder, J.D.; Li, D.; Herron, J.A.; Mavrikakis, M.; Chi, M.; More, K.L.; Li, Y.; Markovic, N.M.; Somorjai, G.A.; Yang, P.; Stamenkovic, V.R. 2014. Highly crystalline multimetallic nanoframes with three-dimensional electrocatalytic surfaces. *Science* 343:1339–1343.
45. Chattot, R.; Le Bacq, O.; Beermann, V.; Kühl, S.; Herranz, J.; Henning, S.; Kühn, L.; Asset, T.; Guétaz, L.; Renou, G.; Drnec, J.; Bordet, P.; Pasturel, A.; Eychmüller, A.; Schmidt, T.J.; Strasser, P.; Dubau, L.; Maillard, F. 2018. Surface distortion as a unifying concept and descriptor in oxygen reduction reaction electrocatalysis. *Nature Materials* 17:827–833.
46. Nørskov, J.K.; Rossmeisl, J.; Logadottir, A.; Lindqvist, L.; Kitchin, J.; Bligaard, T.; Jonsson, H. 2004. Origin of the ove rpotential for oxygen reduction at a fuel-cell cathode. *Journal of Physical Chemistry B* 108:17886–17892.
47. Colic, V.; Bandarenka, A.S. 2016. Pt-Alloy electrocatalysts for the oxygen reduction reaction: from model surfaces to nanostructured systems. *ACS Catalysis* 6:5378–5385.
48. Stephens, I.E.L.; Bondarenko, A.S.; Grønbjerg, U.; Rossmeisl, J.; Chorkendorff, I. 2012. Understanding the electrocatalysis of oxygen reduction on platinum and its alloys. *Energy & Environmental Science* 5:6744–6762.
49. Sabawa, J.; Bandarenka, A.S. 2019. Degradation mechanisms in polymer electrolyte membrane fuel cells caused by freeze-cycles: investigation using electrochemical impedance spectroscopy. *Electrochimica Acta* 311:21–29.

50. Lochner, T.; Hallitzky, L., Perchthaler, M.; Obermaier, M. Sabawa, J.; Enz, S.; Bandarenka, A.S. 2020. Local degradation effects in automotive size membrane electrode assemblies under realistic operating conditions. *Applied Energy* 260: 114291.
51. Dubau, L.; Castanheira, L.; Maillard, F.; Chatenet, M.; Lottin, O.; Maranzana, G.; Dillet, J.; Lamibrac, A.; Perrin, J.C.; Moukheiber, E.; ElKaddouri, A.; De Moor, G.; Bas, C.; Flandin, L.; Caqué, N. 2014. A review of PEM fuel cell durability: materials degradation, local heterogeneities of aging and possible mitigation strategies. *Wiley Interdisciplinary Reviews: Energy and Environment* 3:540–560.
52. Meier, J.C.; Katsounaros, I.; Galeano, C.; Bongard, H.J.; Topalov, A.A.; Kostka, A.; Karschin, A.; Schüth, F.; Mayrhofer, K.J.J. 2012. Stability investigations of electrocatalysts on the nanoscale. *Energy & Environmental Science* 5:9319–9330.
53. Pollet, B.G.; Kocha, S.S.; Staffell, I. 2019. Current status of automotive fuel cells for sustainable transport. *Current Opinion in Electrochemistry* 16:90–95.
54. Batchelor, T.A.A.; Pedersen, J.K.; Winther, S.H.; Castelli, I.E.; Jacobsen, K.W.; Rossmeisl, J. 2019. High-entropy alloys as a discovery platform for electrocatalysis. *Joule* 3:834–845.
55. Lopes, P.P.; Strmcnik, D.; Tripkovic, D.; Connell, P.P.; Stamenkovic, V.; Markovic, N.M. 2016. Relationships between atomic level surface structure and stability/activity of platinum surface atoms in aqueous environments. *ACS Catalysis* 6:2536–2544.
56. Geiger, S.; Kasian, O.; Ledendecker, M.; Pizzutilo, E.; Mingers, A.M.; Fu, W.T.; Diaz-Morales, O.; Li, Z.; Oellers, T.; Fruchter, L.; Ludwig, A.; Mayrhofer, K.J.J.; Koper, M.T.M.; Cherevko, S. 2018. The stability number as a metric for electrocatalyst stability benchmarking. *Nature Catalysis* 7:508–515.

CHAPTER 4

Ionic Conductors

4.1 ELECTROLYTES AND SOME DIFFERENCES BETWEEN LIQUID AND SOLID ION CONDUCTORS

As it was mentioned in previous sections, many energy conversion and storage devices deal with spontaneous spatially separated redox processes. The latter generally require suitable ionic conductors. Naturally, the performance of these devices depends on the properties of the electrode surface, the electrode/electrolyte interface, and of course, the properties of the electrolytes. Liquid and solid ionic conductors are among the most important energy materials.

DEFINITION:

Electrolytes are usually liquids or solids, which conduct electricity through the movement of ions. Ionic conductivity is mutually connected with ion transport under the influence of an external electric field.

There are several properties, which are fundamentally different in the case of electronic and ionic conductors. Among them are the specific conductivity values and their dependencies on temperature (see Table 4.1).

One can arbitrarily distinguish at least three types of ionic conductors: inorganic solid, liquid, and polymer electrolytes. The mechanism of conductivity and the area of applications of these materials are drastically different. Therefore, it is essential to understand what determines the ionic conductivity of solids and liquids, how the nature of electrolyte components influences their functionality, and what the design principles to construct such functional materials are.

DOI: 10.1201/9781003025498-4

TABLE 4.1 Some Differences between Electron and Ionic Conductors

Electron Conductors, e.g., Metals	Ionic Conductors, e.g., Some Ceramic Materials
• Conductivity range: 10 S/cm $<\sigma<10^5$ S/cm	• Conductivity range: 10^{-3} S/cm $<\sigma<10$ S/cm
• Electrons carry the current	• Ions carry the current
• Conductivity decreases linearly as temperature increases (phonon scattering increases with temperature)	• Conductivity increases exponentially with temperature (activated transport)

Why not start with the basic phenomena and properties related to ion conductors. First of all, it is essential to define the typical ways of ion transport in such materials. There are three characteristic modes of ion transport in ionic conductors listed below.

Migration – In this mode, charged species move to equalize gradients of the potential in electrolytes.

Convection – In such a transport mode, the material is forced to move by an external force such as the electrode's rotation. This mode is typical for liquid electrolytes.

Diffusion – In this mode, a species' movement occurs under the influence of a gradient of chemical potential (generally, due to a concentration gradient).

Interestingly, even from the modes of transport listed above, one can probably see that there are functional differences related to ion transport and structural configuration between solid and liquid electrolytes. From one point of view, solid electrolytes would be more desirable in larger-scale energy applications. While solid electrolytes have certain disadvantages, they help to address many issues with system design, corrosion problems, etc. On the other hand, in liquid ion conductors, it is possible to change the electrolyte concentration easily and thus control the ionic conductivity at a given temperature. In the case of liquids, one can also add an inert *supporting electrolyte*, which helps to minimize the effects of nonuniform electric field distribution in the system.

Solid electrolytes can be single-crystalline, amorphous, and polycrystalline. Charges of one sign can, in many cases, be practically immobile in such materials. Another peculiarity is that, in general, liquid electrolytes have negligible electronic conductivity. At the same time, numerous solids show noticeable electronic conductivity even at room temperature. Finally, in solid ion conductors, mobile ions can move as close to an electrode surface as permitted by ion size considerations. In contrast, there is a compact layer composed of solvent molecules (e.g., H_2O) around dissolved ions in liquid electrolytes, which influences the electrolyte properties significantly [1,2]. Therefore, one can expect, e.g., capacitive effects that are considerably different between solid and liquid ion conducting systems.

4.2 CONDUCTIVITY OF LIQUID ELECTROLYTE SOLUTIONS

The high ionic conductivity of liquids often originates from the fact that certain substances can dissociate in liquid solvents (exceptions are the ionic liquids (ILs) discussed later), forming solvated anions and cations. Typically, both of them contribute to the resulting conductivity due to their mobility. For example, molecules of sulfuric acid, H_2SO_4, in the presence of water dissociate according to the following scheme:

$$H_2SO_4 \leftrightarrow H^+ + HSO_4^{2-}$$

$$HSO_{4^-} \leftrightarrow H^+ + SO_4^{2-}$$

One should note here that concentrated aqueous solutions of sulfuric acid are, for instance, critical electrolytes in some widely used types of batteries (e.g., in lead-acid batteries, as will be discussed later). The high ionic conductivity of such aqueous solutions, among other properties, has essentially contributed to the overall success of these devices' applications.

The mass transport of ions in liquids is much facilitated compared to solids; one can vary the ion conductivity by merely changing the concentration of the dissociating species. However, it is essential to note a rather complex dependence of the ionic conductivity on electrolyte concentration. For instance, it is not possible to continuously increase the conductivity of aqueous solutions of H_2SO_4 by increasing their concentration. At some point, it goes down, as schematically shown in Figure 4.1. The explanation of such a phenomenon, however, is relatively simple. With the increase in the concentration, the fraction of nondissociated H_2SO_4 molecules increases, as there is not enough solvent to make all the molecules dissociate fully. Thus, the number of ions responsible for the conductivity and the conductivity itself decrease.

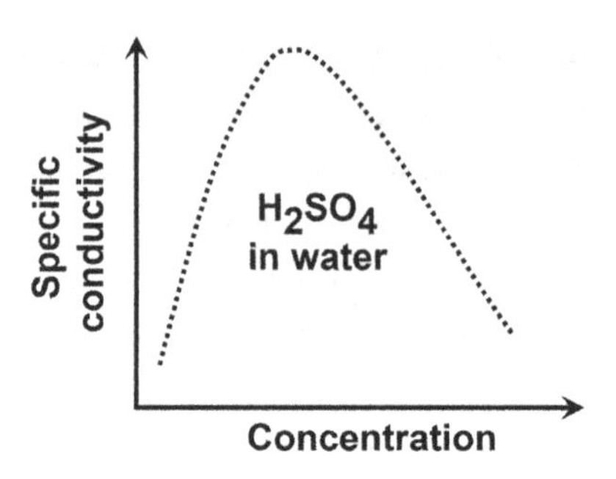

FIGURE 4.1 A schematic dependence of the ionic conductivity of aqueous solutions of sulfuric acid on its concentration.

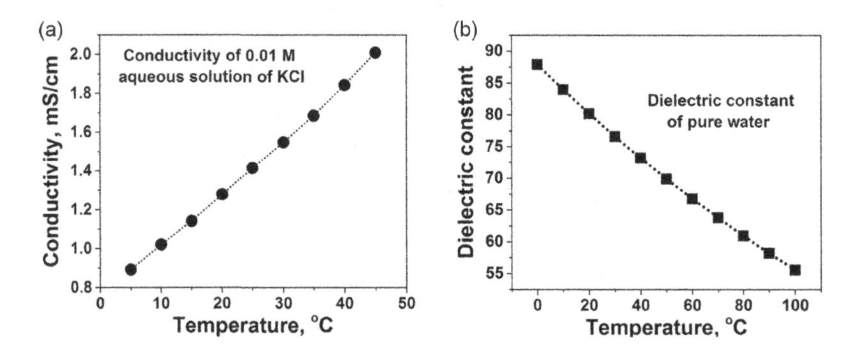

FIGURE 4.2 (a) Ionic conductivity of an aqueous 0.01 M solution of potassium chloride as a function of temperature. (b) Temperature dependence of the dielectric constant of pure water.

The conductivity of liquid electrolytes often increases with increasing temperature. However, it is often difficult to predict such dependencies, as many other parameters, for example, the degree of dissociation or the dielectric constant of the solvent, noticeably change upon heating. Figure 4.2a shows one relatively simple example related to the temperature dependence of the conductivity of an aqueous solution of KCl. In this case, one can see that the ionic conductivity gradually increases if the temperature is increased. Simultaneously, the dielectric constant of pure water decreases, changing polarizability and, therefore, the solvent's ability to form solvated anions or cations (Figure 4.2b).

4.3 IONIC CONDUCTIVITY OF SOLID ELECTROLYTES

The high ionic conductivity observed in solids frequently depends upon fundamentally dissimilar phenomena and parameters. One can distinguish different conductivity mechanisms in, e.g., oxides, the so-called superionic solid salts, and other materials, like polymers. The state-of-the-art understanding of this kind of materials gives an idea that ionic conductivity of solid electrolytes acceptable for energy applications can originate from the following:

i. The number and nature of the so-called point, linear, and planar defects

ii. A unique phenomenon, which can be roughly described as a "corporative behavior" of specific structural units in solids facilitating fast ion transport in *superionic phases*

iii. Superstructures consisting of highly conductive and nonconductive fragments (this often happens in ionically conducting polymers). The nonconducting fragments are responsible for the mechanical stability and stability against degradation caused by external physical and chemical factors, such as high-energy radiation or exposure to reactive substances

In the following, the cases mentioned above will be considered one by one.

For case number one, imagine a single crystal of a solid compound. Somewhat surprisingly, classical thermodynamics predicts that almost any single crystal should have some bulk defects. Figure 4.3 schematically demonstrates the impact of enthalpy and entropy factors on the system's resulting free energy.

With some degree of simplification, both factors influence the system differently, leading to a minimum in free energy at a certain number (concentration) of defects. In simple terms, if one starts with a hypothetically perfect crystal, external energy should be spent to introduce imperfections. However, the more the defects are created, the higher the number of possible configurations of structural units is possible inside the solid (configurational entropy $S = k \ln W$, where W is the number of possible configurations, and k is the Boltzmann constant). When the concentration of defects is small, the entropy factor dominates. However, when the defect concentration is high enough, new defects do not change the entropy significantly. At this point, the enthalpy factor starts playing a more significant role. Recalling the famous Gibbs equation $\Delta G = \Delta H - T\Delta S$, one can conclude that nature always wants to create some degree of disorder even in "perfect systems" (Figure 4.3) [3].

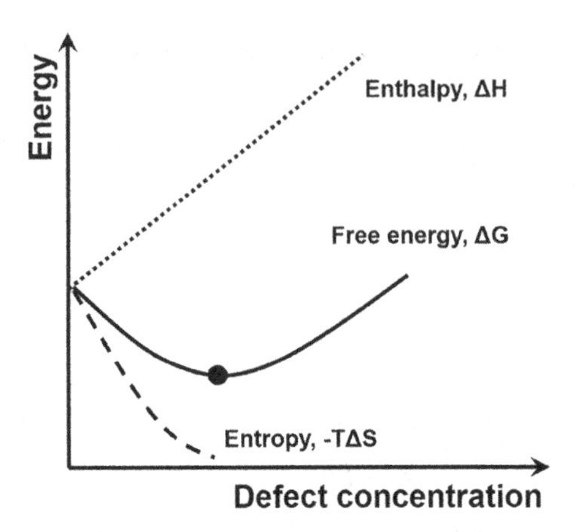

FIGURE 4.3 Energy changes upon introducing defects into a perfect crystal. Configurational entropy is given as $S=k \ln W$, where W is the number of possible configurations, and k is the Boltzmann constant. The free energy is given as $\Delta G=\Delta H-T\Delta S$. Note the minimum in the free energy at a certain concentration of defects (marked with a dot).

The idea that the high ionic conductivity of some solids can be due to the defects in the crystal structure probably originates from the work by W. Nernst published in 1899 [4]. It has been elaborated further during the 20th century and created a basis of the current understanding of this phenomenon [5–9]. Suppose the ionic conductivity of solids is determined entirely by defects. In that case, one can distinguish at least two simple types of *point defects* contributing to the resulting conductivity: Schottky and Frenkel defects (see Figure 4.4) [3,10–12]. If one considers the Schottky defects, this type of imperfections is nothing more than missing structural units (Figure 4.4a), such as cations or anions.

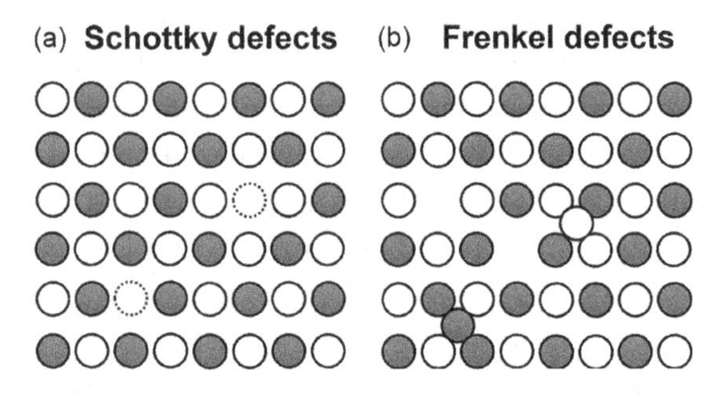

FIGURE 4.4 Schematic representation of (a) Schottky and (b) Frenkel defects in solid ionic conductors.

The Frenkel defects designate random structural units between regular crystal structure units (Figure 4.4b).

The other types of defects are *line defects*. Those appear mostly as a result of misalignment of the structural units like ions or arise in the presence of vacancies "along a line". This kind of defects is sometimes difficult to imagine, and Figure 4.5 is intended to be helpful in creating the first impression of a possible defective structure in this case.

The third type of defects essential for ionic conductivity is *planar defects*. Those defects are, for example, grain boundaries, as schematically shown in Figure 4.6. It is probably not straightforward to understand the nature of these defects in the context of their contribution to the overall ionic conductivity of solids. They can be considered as interfaces between closely packed crystallites in a polycrystalline material. The first realistic quantitative models dealing with the assessment of their influence on the overall conductivity appeared only in the 1960s [13]. Other breakthrough ideas pointed out that the grain boundaries must be "highways" for the ions, mainly contributing to the ion conductivity, if there is no segregation of impurities at those boundaries [14–16]. The grain boundaries are the places where the ion mobility must, in many cases, be the highest in the most defect-rich directions, i.e., at the interface [16].

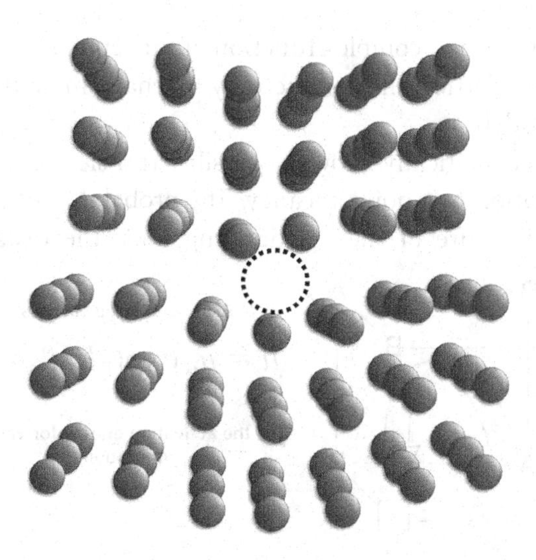

FIGURE 4.5 A model, which visualizes the so-called line defects. This kind of defects results from missing ions in crystal structures along "a line" (dotted circle designates a missing row).

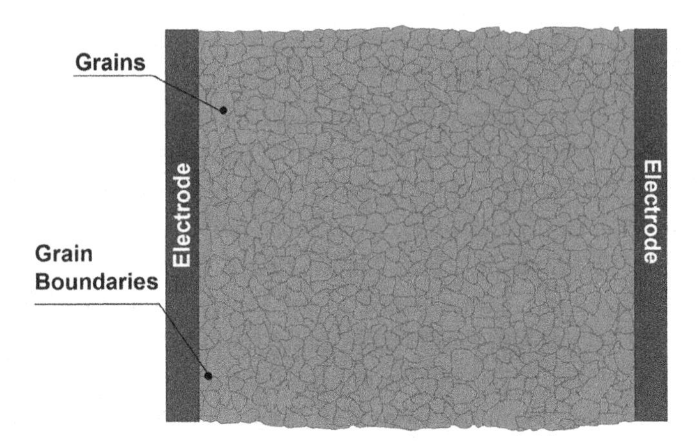

FIGURE 4.6 Grain boundaries are planar defects, which play a crucial role in various polycrystalline ionic conductors. They can be considered as interfaces between closely packed crystallites in a solid electrolyte.

At this point, one can start quantifying the ionic conductivity due to defects, σ, with the observations that it is directly proportional to the number of charge carriers (i.e., ions), n, their charge, e, and their mobility, μ (see Figure 4.7):

$$\sigma = n \cdot e \cdot \mu$$

The mobility of ions is a complex function of the temperature, the properties of the crystal lattice, the conductivity mechanism, and the nature of the dominating defects.

Figure 4.7 schematically shows a possible transfer of an ion A from its original position to a point vacancy. The probability of this jump will depend on the nature of the surrounding ions, the distance between

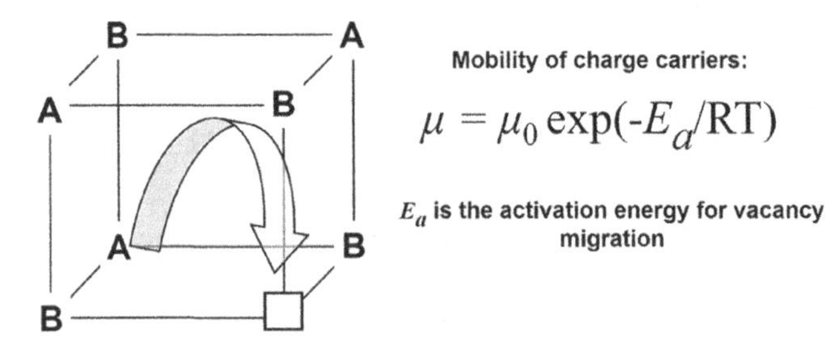

FIGURE 4.7 Defects often enable high ionic conductivities. The resulting ionic conductivity depends on an activation barrier to move ions from their initial stable positions toward a vacancy, as shown in the picture.

initial and final points, and of course, the parameters of the crystal lattice. It will also depend on surrounding defects and the crystal temperature, T. Experiments show that the species' mobility can be expressed in the form presented in Figure 4.7. It is an exponential function of temperature (Arrhenius type of behavior). The so-called activation energy, E_a, quantifies the energy barrier to overcome for an ion to move inside solid electrolytes *via* the mechanism involving defects. Overall, the dependences of the ionic conductivity of solid electrolytes on temperature can be expressed with the following formula:

$$\sigma = \sigma_0 \exp(-E_a / RT)$$

where R is the universal gas constant (~8.31 $J{\cdot}K^{-1}{\cdot}mol^{-1}$).

Consequently, if one plots the logarithm of ionic conductivity as a function of the inverse temperature, a linear dependence should be observed. One example of such behavior is related to the O^{2-} conducting electrolyte, $Zr_{0.9}Y_{0.1}O_{1.95}$, used in solid oxide fuel cells. The corresponding conductivity dependence is shown in Figure 4.8. The slope of this kind of dependence can be used to calculate the value of the activation energy. Experimental measurement of the temperature dependence of the ionic conductivities

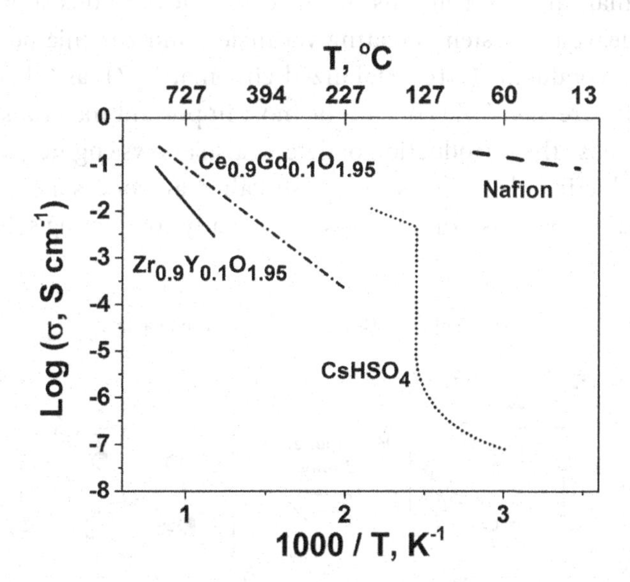

FIGURE 4.8 The conductivity of state-of-the-art solid electrolytes over a wide temperature range. Note the gap in highly conducting materials in the temperature range ~100°C–500°C. (Adapted from [17].)

is a relatively easy method to obtain the basic properties of numerous ion-conducting electrolytes.

4.4 HOMOGENEOUS AND HETEROGENEOUS DOPING OF SOLID ELECTROLYTES

How can one increase the ionic conductivity of solids? The effect depends on the mechanism of the conductivity. If it entirely originates from the crystal structure defects, the main goal is to maximize their amount while maintaining the crystal's stability. One can start considerations with the so-called homogeneous doping [18], which has the particular idea to modify solid electrolytes' chemical composition to create vacancies in the structure.

Consider a hypothetically perfect crystal of zirconium dioxide (zirconia), ZrO_2 (Figure 4.9, left). It is known that in this crystal, oxygen ions are quite mobile. Therefore, one can increase the ionic conductivity of this material by creating the maximum number of oxygen vacancies. One way to create such a situation is to replace Zr^{4+} ions in the lattice with cations with a reduced effective charge. It is possible to realize this approach using, for example, Y^{3+} cations as the doping species. To keep electroneutrality, one should eliminate the excessive negative charge in the crystal caused by the "remaining" oxygen ions in this case. These anions should spontaneously leave the system, creating vacancies, and zirconia becomes an oxygen ion conductor (yttria-stabilized zirconia, YSZ), as schematically shown in Figure 4.9. YSZ is one of the most important materials for solid oxide fuel cells. The introduction of defects also allows engineering highly Li-ion conducting electrolytes for the so-called all-solid-state Li-ion batteries [19–22] as well as other numerous ionically conducting solids.

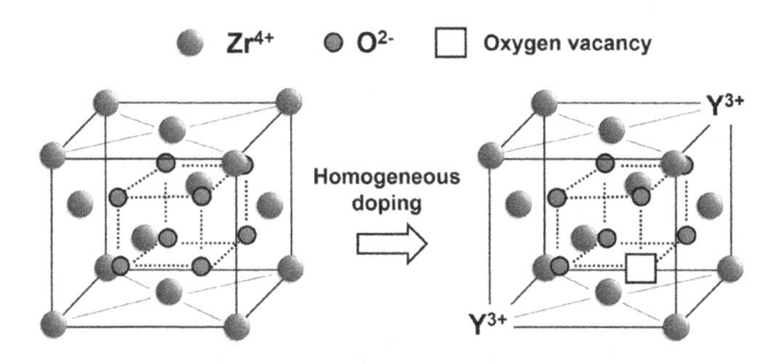

FIGURE 4.9 Schematics explaining the idea of homogeneous doping using yttria-stabilized zirconia as an example.

Homogeneous doping has numerous advantages. However, it is often complicated to end up with a structure and composition, which are full of defects and at the same time stable, preventing the formation of separate phases with a reduced number of vacancies.

From the description given above, one can conclude that for oxygen species to leave the crystal (normally, this is facilitated upon heating), a certain partial pressure of molecular oxygen, pO_2, in the crystal's surroundings is necessary. The exact mechanisms of how the system can keep electroneutrality are quite complex. Naturally, the ionic O^{2-} conductivity of solids depends on the partial pressure of oxygen, pO_2. While there are parameter sets where the ionic conductivity is stable irrespective of the O_2 content around the crystal, at some point, it changes with pO_2. One illustrative example is given in Figure 4.10 for the case of $LaYO_3$ ceramics. This material's conductivity gradually increases with temperature, but it is also very dependent on $\log(pO_2)$, with an almost constant slope, indicating the same type of conductivity mechanisms.

Another approach to increase the overall ionic conductivity of solids involves the so-called heterogeneous doping [18,24]. It is essential to notice that this approach is again mainly valid for solid electrolytes, where only defects in the crystal structure are responsible for the resulting conductivity. The idea behind heterogeneous doping might appear a bit counterintuitive. In this case, the classical approach is to introduce insulating species into the bulk of the ion conductor. While a dielectric replaces a sacrificial volume, one creates some amount of material between the added insulator particles and the conducting phase, which is full of different types of

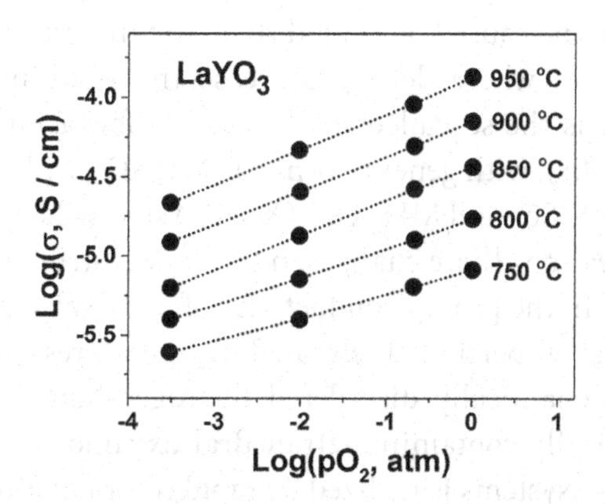

FIGURE 4.10 Dependences of the ionic conductivity of $LaYO_3$ on oxygen partial pressure at various temperatures. (The graph is adapted from [23].)

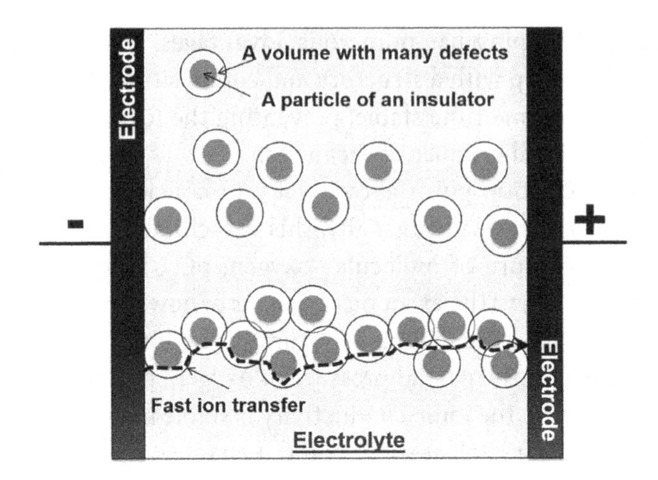

FIGURE 4.11 Heterogeneous doping to increase the ionic conductivity of solids. Some fraction of particles of an insulator is added to create interconnected disordered areas between the electrodes. Ion transport is facilitated in such areas. A dashed arrow schematically shows the fastest way for the ions.

defects. It is important to realize that it is essential to have some particle content when the areas containing the high content of defects overlap each other and create continuous passways between electrodes. Figure 4.11 explains that it is crucial to have a continuous overlap of the defect-rich areas to compensate for the replacement of ion-conducting volumes with a dielectric phase.

4.5 SUPERIONIC CONDUCTORS

So far, only one ionic conductivity mechanism of solids has been considered, namely the one caused by crystal structure imperfections. However, there are materials where defects decrease the resulting conductivity. A good example is the so-called *solid acids* – a few acidic salts, such as $CsHSO_4$ (Figure 4.8), with general formulas $MHXO_4$, $M_3H(XO_4)_2$ (M=Cs, Rb, NH_4 and M=S, Se), and $MH_2X'O_4$ (X'=P, As), which exhibit structural changes upon heating. These changes in the crystal structure lead to a significant increase in the proton conductivity of these salts [25–28].

Superprotonic properties at elevated temperatures result from the existence of a dynamically disordered hydrogen-bond network in the compounds typically containing tetrahedral oxyanions. The fast proton transport in these systems is realized by rapid re-orientation of some specific structural units, such as the XO_4 tetrahedra in the partially disordered structure of, e.g., $CsHSO_4$.

Figure 4.12 shows a schematic dependence of the proton conductivity of solid acids on the inverse of the temperature. One can typically observe a jump in the conductivity when structural changes occur in the system. The ionic conductivity before the superionic phase transition is cased entirely due to the number and nature of defects in the material. After the phase transition, the increased conductivity is mainly due to individual structural units' corporative behavior in the solids. Therefore, the concepts of homogeneous and heterogeneous doping usually do not work in the case of these systems at elevated temperatures: the more defects one introduces, the higher the conductivity of the low-temperature phase. However, the conductivity becomes lower in this case at temperatures higher than the superionic phase transition. Consequently, it is essential to have a pure solid acid phase to enable the highest possible proton conductivity.

From the considerations presented above, it is also clear that solid acids (single crystals) should demonstrate anisotropy of the proton conductivity in different directions. Figure 4.13 schematically explains that proton transport is largely facilitated only in one direction, where the rotational degree of freedom is enabled.

Nowadays, solid acids are used in the so-called *solid acid fuel cells* [29], which are commercially available and operate at elevated temperatures (*ca* 200°C–300°C). One can use cheaper nonnoble catalysts at these temperatures, and the resulting current densities are reasonable for many applications, including automotive ones. However, the poor mechanical stability of thin-film electrolytes is a big drawback of solid acids. Additionally, wider applications meet challenges because these electrolytes often use

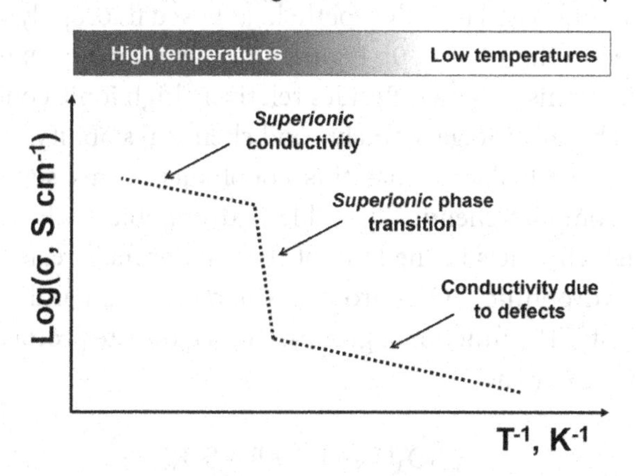

FIGURE 4.12 A schematic dependence of the proton conductivity of solid acids as a function of temperature.

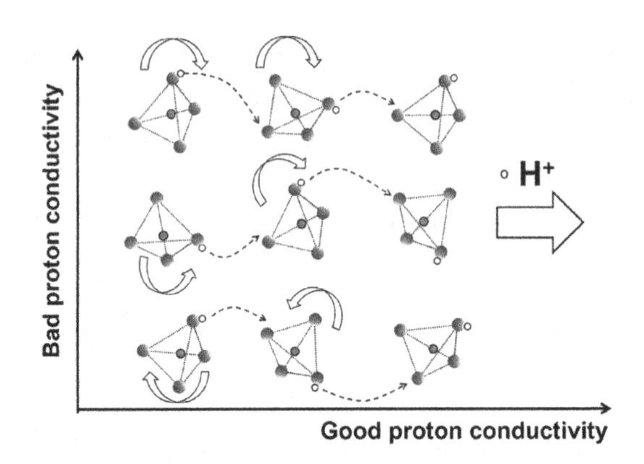

FIGURE 4.13 Schematics demonstrating anisotropy of the proton conductivity in superionic solid acids. The rotational degree of freedom in the structure is enabled only along a specific direction in single crystals: H^+ transport is fast only along those routes.

scarce Cs, and new affordable, highly conducting, and stable solid acid electrolytes are necessary.

4.6 POLYMER ION CONDUCTORS

Polymer ion-conducting membranes are materials of choice for numerous energy applications, as they usually provide flexibility in the system design and manufacturing schemes. Probably the most important and widely used polymer of this kind is the proton-conducting "Nafion". This name appeared first as a trademark. However, now it is commonly used in scientific literature as the polymer's distinguishable name.

Nafion is a sulfonated tetrafluoroethylene-based fluoropolymer-copolymer developed in the 1960s [30]. Figure 4.14 shows an approximate structural formula of this polymer. Besides relatively high ionic conductivity, it has a remarkable advantage, namely good chemical stability.

What is also valuable is that this copolymer forms superstructures [31–33] that combine chemically stable hydrophobic $[-CF_2-]_x$ fragments and water-rich channels in the bulk of the membrane, created due to the presence of hydrophilic $-SO_3H$ groups, which give high and durable proton conductivity. The functional groups, $-SO_3H$, donate protons according to the following scheme:

$$R-SO_3H \leftrightarrow H^+ + R-SO_3^-$$

Where R- is used to designate the rest of the polymer.

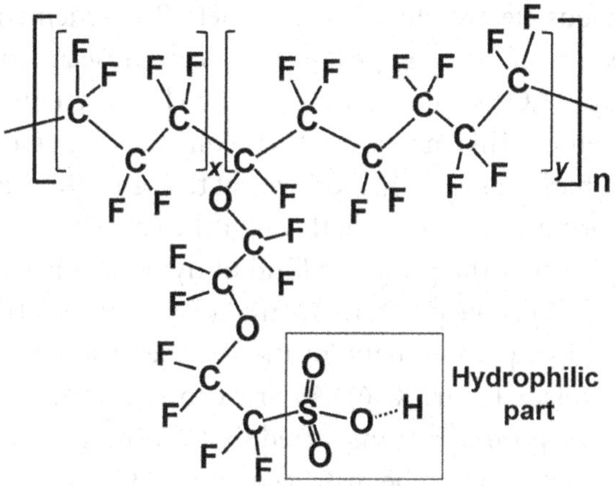

FIGURE 4.14 The approximate structural formula of Nafion polymer.

Figure 4.15a gives a schematic overview of the fragment of the Nafion superstructure. An approximate 3D overview is presented in that figure, explaining how water channels can form stable areas inside the polymer providing high proton conductivity. Figure 4.15b, in turn, shows the model of the channels, where only the polymer chains with the -SO_3H groups are shown (no water molecules are depicted) for better visual clarity.

It is essential to realize that the proton conductivity mechanism in Nafion is not due to the direct jumps of H^+ species from one - SO_3^- functional group to another [34]. The -SO_3H fragments only release protons to the channels filled in with water molecules and thus form an acidic aqueous electrolyte there. In such a medium, the transport of H^+ is relatively fast due to the so-called Grotthuss mechanism. This mechanism explains how the "excess" proton moves through the hydrogen bond network in

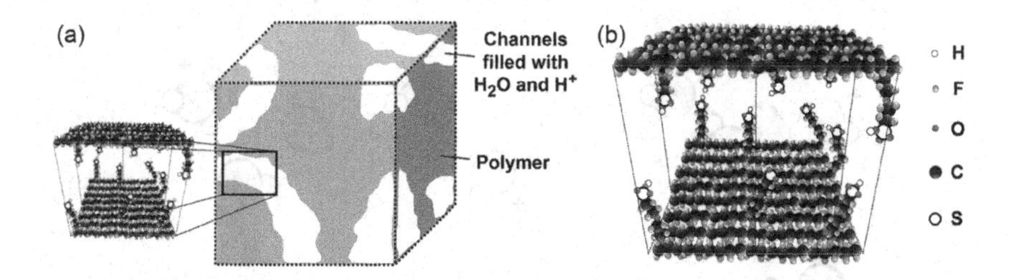

FIGURE 4.15 A model, which describes polymer-water superstructures in the Nafion proton-conducting polymer. (a) A general overview of polymer clusters and water channels. (b) A schematic of a channel with hydrophobic and hydrophilic –SO_3H fragments (water molecules in the channel are not shown for clarity).

the acidic aqueous electrolyte in the channels. The schematic illustrating this mechanism is given in Figure 4.16. As can be seen from the figure, there are always H_3O^+ species present in the electrolyte in the channels, not isolated protons. This means the H^+ is continuously associated with at least one water molecule [35,36]. However, at the same time, it is connected with a neighboring H_2O molecule through the so-called hydrogen bond. One can imagine that the proton is kind of dynamically shared between two H_2O species. At some point, there might be a fast redistribution of the bonds resulting in a predominant location of the proton in the neighboring H_2O species (see Figure 4.16). As one can see from the figure, not a single "excessive" proton is transported, but an indirect transport takes place *via* a redistribution of the hydrogen bonds [37].

It is clear from the mechanism described above that evaporation of water from the Nafion polymer superstructure would decrease the proton conductivity of the polymer electrolyte membrane. Indeed, Figure 4.17 shows that the conductivity of Nafion significantly decreases even at temperatures below 100°C due to water evaporation. Therefore, constant humidification of Nafion is required even at lower temperatures. This kind of behavior largely determines the performance of some working devices, e.g., certain fuel cells or electrolyzers. Even under 75% relative humidity conditions, the operation is limited by *ca* 80°C. The high proton conductivity of Nafion existing only at relatively low temperatures (and acidic media) limits the choice of electrocatalysts in electrolyzers and fuel cells, in which this electrolyte is used. Nevertheless, Nafion is still a key material in the polymer electrolyte fuel cells and electrolyzers considered in the previous chapter.

Another drawback of Nafion is that it is relatively expensive to synthesize, while still noticeable degradation of this polymer with time is

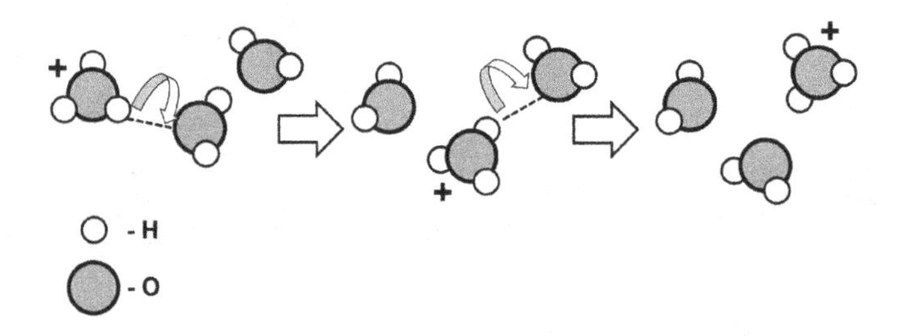

FIGURE 4.16 Schematic representation of the proton transport according to the Grotthuss mechanism in aqueous electrolytes. The fast proton transfer is due to the hydrogen bond network dynamics.

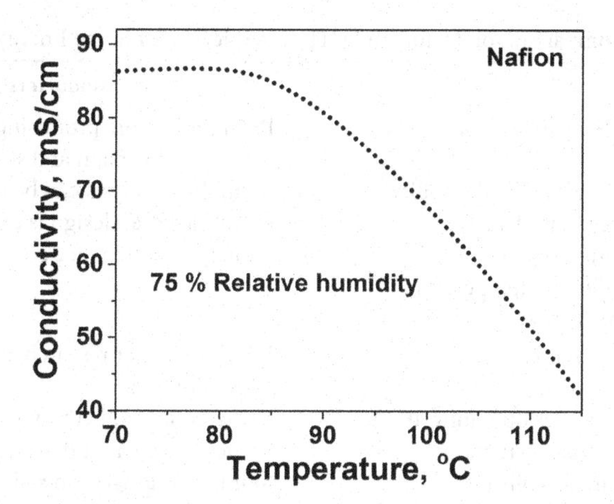

FIGURE 4.17 Typical dependence of the proton conductivity on temperature for the Nafion polymer under humidification (75% relative humidity).

also observed. All these issues motivate the research community to find more affordable and stable alternatives [38,39]. The design principles in these attempts remain, however, the same. One should design a polymer consisting of some hydrophobic polymer chains with hydrophilic -SO_3H functional groups. Those would then create chemically and mechanically stable superstructures with the channels filled with aqueous acidic electrolytes. One example of such a polymer is sulfonated polyether ether ketone, sPEEK (see Figure 4.18).

While sPEEK is synthesized according to the same design rules as in the case of Nafion, the performance of the resulting membranes for energy applications is often not as good compared to the Nafion polymer.

The development of chemically and mechanically stable membranes, which have high OH^- conductivity, is also of high priority for energy applications. In this case, low-cost Ni-, Co-, and Fe-based electrocatalysts can be used at both cathode and anode sides of low-temperature electrolyzers

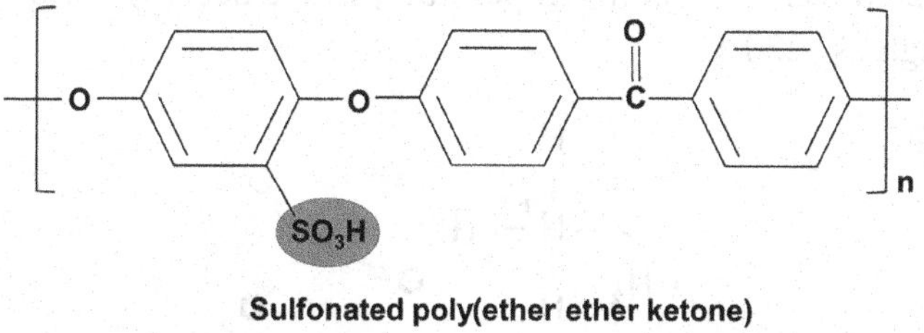

Sulfonated poly(ether ether ketone)

FIGURE 4.18 The approximate structural formula of sulfonated polyether ether ketone. Note that the functional –SO_3H group is the same as in Nafion polymer.

TABLE 4.2 Comparison of Proton and OH^- Conductors for Some Energy Applications

H^+ conductors	OH^- conductors
The materials are ready for commercial applications: provide reasonable chemical stability provide higher current densities have higher conductivities have wider operational temperature window for applications	Potentially more promising for many energy conversion and storage devices: nonnoble catalysts can be used more "flexible" design is possible reduced maintenance costs are envisaged
Require further research and development: to develop membranes operating at "intermediate" temperatures to increase thermal stability to improve long-term and mechanical stability	Require further research and development: to increase conductivities to increase current densities to improve long-term stability

and fuel cells instead of Pt-based and Ir-based ones. Table 4.2 compares the advantages and possible issues for the proton and hydroxyl conducting membranes, as discussed above.

Recently, there has been noticeable progress in designing OH-conducting polymers [40–44], and one can foresee that stable and highly conducting membranes can be commercialized in the near future.

4.7 IONIC LIQUIDS

DEFINITION:

An IL is a salt, which is in the liquid state at temperatures, arbitrarily selected below ~100°C (ideally, it is a liquid already at room temperature). Consequently, ILs are liquids composed of ions only. In principle, any salt that melts without decomposing or vaporizing usually yields an IL at some temperature.

The earliest (1914) reported example of a room-temperature IL was ethyl ammonium nitrate (see Figure 4.19). Its ionic conductivity was reported to be about 20 mS·cm^{-1} at room temperature, with a decomposition temperature as high as ~250°C.

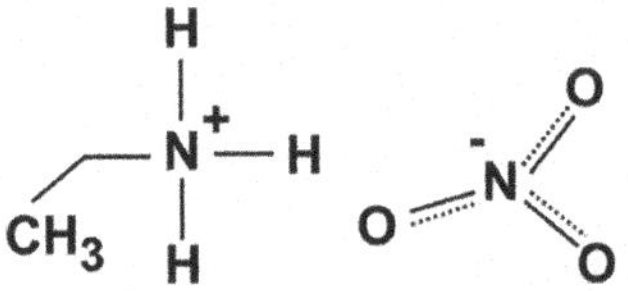

FIGURE 4.19 Structural formulas of the cation and anion of the first reported IL, ethyl ammonium nitrate.

ILs found various areas of applications ranging from advanced batteries [45], dye-sensitized solar cells [46], double-layer supercapacitors [47], and actuators to fuel cells, thermocells, etc. [48]. They are, in many cases, nonflammable. For instance, the replacement of the conventional flammable and volatile organic solutions with ionic-liquid-based, lithium-ion conducting electrolytes may significantly reduce, if not prevent, the risk of thermal runaway of traditional Li-ion batteries (will be considered later). There were also reports that specific ILs completely changed electrocatalysts' properties, resulting in reduced overpotentials or alterations in the catalyst selectivity [49–52].

A low melting point is one of the most important parameters for ILs. On the one hand, it is essential to increase the effective size of the cations and anions to weaken interactions between effective positive and negative charges to decrease the melting point. On the other hand, by varying the chemical nature of the cations and anions, one can also tailor the desired chemical properties of the resulting ILs (e.g., produce aprotic, protic, and/or zwitterionic ones) and extend their stability window against electrochemical decomposition far beyond that of water. Some common structural units, i.e., cations and anions, to design ILs for energy applications are given in Figure 4.20.

4.8 SUMMARY AND CONCLUSIONS

A wide range of ion conductors attracts growing attention as energy materials. There are different classes of ion conductors ranging from common liquid aqueous electrolytes, proton and O^{2-} conducting solid materials, H^+ and hydroxyl conducting polymer membranes to ILs. However, challenges in their design and development remain the same: to increase the conductivity, improve the mechanical properties, and extend the lifetime.

Proton conductors are important for electrolyzers, fuel cells for automotive applications, and other types of energy conversion and storage devices, such as supercapacitors or batteries. Great efforts are dedicated to developing new low- and intermediate-temperature solid proton and especially

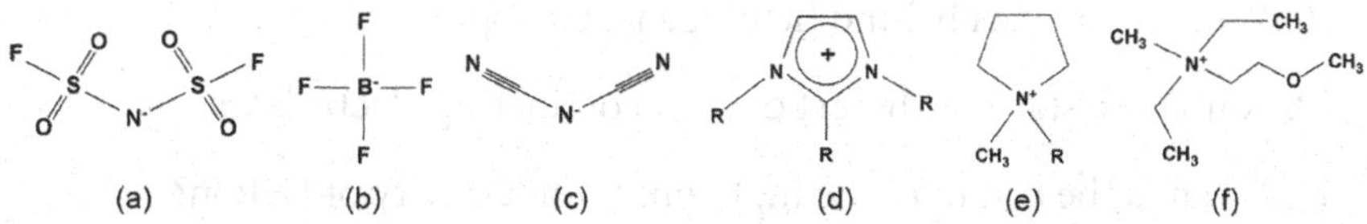

(a) (b) (c) (d) (e) (f)

FIGURE 4.20 Examples of building blocks (anions and cations) of ILs with relatively low melting points. (Adapted from [53].)

polymer ion conductors. Recently, significant attention is also attracted to creating OH-conducting membranes for fuel cells and electrolyzers to enable the use of more affordable electrocatalysts. Commercial Nafion polymer still dominates the market of proton-conducting membranes.

Oxygen-conducting ceramics find most applications in solid oxide fuel cells. In this case, an additional target is to decrease the operational temperature (increase the conductivity at lower temperatures), simplifying the design and use of these devices. There are two main ways to do so: to apply homogeneous and heterogeneous doping or to find an entirely new class of highly conducting ceramics.

ILs can be ideal candidates to play a more important role in substituting solvents in, e.g., Li-ion batteries or supercapacitors. However, more research is necessary to identify affordable and suitable anion-cation combinations with a wide stability window and high conductivity at the current stage.

4.9 QUESTIONS

1. What are ionic conductors?
2. What is the role of ion conductors in energy conversion and storage devices?
3. Analyze the differences between ionic and electronic conductors.
4. What are the main modes of ion transport in ionic conductors?
5. What are the main differences between solid and liquid electrolytes used in energy conversion and storage?
6. What are the typical dependencies of ionic conductivity of liquids and solids on temperature?
7. What is the typical origin of the high ionic conductivity of solids?
8. What is the idea behind homogeneous doping?
9. What is the idea behind heterogeneous doping?
10. Name the state-of-the-art oxygen conducting electrolytes.
11. What is the origin of the high ionic conductivity of Nafion?
12. Explain what the Grotthuss mechanism of proton transport is.

13. What are the main design principles of new ion conductors?
14. What are the "superionic" phase transitions? Name the state-of-the-art proton-conducting "solid acids". Analyze the pros and cons of the OH^- and H^+ solid ion conductors.
15. Define ILs.
16. What are the design principles of ILs with low-temperature melting points?

REFERENCES

1. Barthel, J.M.G.; Krienke, H.; Kunz, W. 1998. *Physical Chemistry of Electrolyte Solutions: Modern Aspects* (Topics in Physical Chemistry, Vol. 5). Steinkopff Verlag: Heidelberg.
2. Andreev, M.; de Pablo, J.J.; Chremos, A.; Douglas, J.F. 2018. Influence of ion solvation on the properties of electrolyte solutions. *The Journal of Physical Chemistry B* 122:14.
3. Wagner, K.W.; Schottky, W. 1930. Theory of controlled mixed phases. *Zeitschrift für Physikalische Chemie*, 11B:163.
4. Nernst, W. 1899. Über die elektrolytische Leitung fester Körper bei sehr hohen Temperaturen. *Zeitschrift für Elektrochemie* 6:41–43.
5. Hund, F. 1951. Anomale Mischkristalle im System ZrO_2-Y_2O_3. Kristallbau der Nernst-stifte. *Zeitschrift für Elektrochemie und Angewandte Physikalische Chemie* 55:363–366.
6. Strickler, D.W.; Carlson, W.G. 1964. Ionic conductivity of cubic solid solutions in the system CaO-Y_2O_3-ZrO_2. *Journal of the American Ceramic Society* 47:122–127.
7. Dixon, J.M.; LaGrange, L.D.; Merten, U.; Miller, C.F.; Porter, J.T. 1963. Electrical resistivity of stabilized zirconia at elevated temperatures. *Journal of the Electrochemical Society* 110:276–280.
8. Maier, J. 2003. Defect chemistry and ion transport in nanostructured materials. Aspects of nanoionics. Part II. *Solid State Ionics* 157:327–334.
9. Funke, K. 2013. Solid state ionics: From Michael Faraday to green energy—the European dimension. *Science and Technology of Advanced Materials* 14:043502.
10. Frenkel J. 1926. Über die Wärmebewegung in festen und flüssigen Körpern. *Zeitschrift für Physik* 35:652–669.
11. Schottky, W. 1935. Über den Mechanismus der Ionenbewegung in festen Elektrolyten. *Zeitschrift für Physikalische Chemie* B 29:335–355.
12. West, A.R. 2014. *Solid State Chemistry and Its Applications*. 2nd Edition, John Wiley & Sons: New York, 584p.
13. Bauerle, J.E. 1969. Study of solid electrolyte polarization by a complex admittance method. *Journal of Physics and Chemistry of Solids* 30:2657–2670.

14. Maier, J. 1986. On the conductivity of polycrystalline materials. *Berichte der Bunsengesellschaft für Physikalische Chemie* 90:26–33.
15. Nafe, H. 1984. Ionic conductivity of ThO_2- and ZrO_2-based electrolytes between 300 and 2000 K. *Solid State Ionics* 13:255–263.
16. Fleig, J. 2003. Microelectrodes in solid state ionics. *Solid State Ionics* 161:279–289.
17. Karlsson, M. 2013. Perspectives of neutron scattering on proton conducting oxides. *Dalton Transactions* 42:317–329.
18. Maier, J. 2017. Doping strategies in inorganic and organic materials. *ZAAC, Journal of Inorganic and General Chemistry* 643:2083–2087.
19. Gautam, A.; Sadowski, M.; Ghidiu, M.; Minafra, N.; Senyshyn, A.; Albe, K.; Zeier, W.G. 2021. Engineering the site-disorder and lithium distribution in the lithium superionic argyrodite Li_6PS_5Br. *Advanced Energy Materials* 11:2003369.
20. Bachman, J.C.; Muy, S.; Grimaud, A.; Chang, H.H.; Pour, N.; Lux, S.F.; Paschos, O.; Maglia, F.; Lupart, S.; Lamp, P.; Giordano, L.; Shao-Horn, Y. 2016. Inorganic solid-state electrolytes for lithium batteries: Mechanisms and properties governing ion conduction. *Chemical Reviews* 116:140–162.
21. Tong, X.; Thangadurai, V.; Wachsman, E.D. 2015. Highly conductive Li garnets by a multielement doping strategy. *Inorganic Chemistry* 54:3600–3607.
22. Zhou, L.; Minafra, N.; Zeier, W.G.; Nazar, L.F. 2021. Innovative approaches to Li-argyrodite solid electrolytes for all-solid-state lithium batteries. *Accounts of Chemical Research* 54:2717–2728.
23. Yamamura, H.; Yamazaki, K.; Kakinuma, K.; Nomura, K. 2002. The relationship between crystal structure and electrical conductivity in the $LaY_{1-x}In_xO_3$ (x=0.0–0.7) system. *Solid State Ionics* 150:255–261.
24. Maier, J. 1986. On the heterogeneous doping of ionic conductors. *Solid State Ionics* 18–19:1141–1145.
25. Sanghvi, S.; Haile, S.M. 2020. Crystal structure, conductivity, and phase stability of $Cs_3(H_{1.5}PO_4)_2$ under controlled humidity. *Solid State Ionics* 349:115291.
26. Ponomareva, V.G.; Bagryantseva, I.N. 2017. Proton conductivity, structural and thermal properties of $(1-x)CsH_2PO_4$-$xBa(H_2PO_4)_2$. *Physics of the Solid State* 59:1829–1835.
27. Bohn, A.; Melzer, R.; Sonntag, R.; Lechner, R.E.; Schuck, G.; Langer, K. 1995. Structural study of the high and low temperature phases of the proton conductor $Rb_3H(SeO_4)_2$. *Solid State Ionics* 77:111–117.
28. Haile, S.M.; Chisholm, C.R.I.; Sasaki, K.; Boysen, D.A.; Uda, T. 2007. Solid acid proton conductors: From laboratory curiosities to fuel cell electrolytes. *Faraday Discussions* 134:17–39.
29. Haile, S.; Boysen, D.; Chisholm, C.; Chisholm, C.R.I.; Merle, R.B. 2001. Solid acids as fuel cell electrolytes. *Nature* 410:910–913.
30. Mauritz, K.A.; Moore, R.B. 2004. State of understanding of Nafion. *Chemical Reviews* 104:4535–4586.

31. Peltonen, A.; Etula, J.; Seitsonen, J.; Engelhardt, P.; Laurila, T. 2021. Three-dimensional fine structure of nanometer-scale Nafion thin films. *ACS Applied Polymer Materials* 3:1078–1086.
32. Dong, B.; Gwee, L.; Salas-de la Cruz, D.; Winey, K.I.; Elabd, Y.A. 2010. Super proton conductive high-purity Nafion nanofibers. *Nano Letters* 10:3785–3790.
33. Yin, C.; Li, J.; Zhou, Y.; Zhang, H.; Fang, P.; He, C. 2018. Enhancement in proton conductivity and thermal stability in Nafion membranes induced by incorporation of sulfonated carbon nanotubes. *ACS Applied Materials and Interfaces* 10:14026–14035.
34. Hiesgen, R.; Morawietz, T.; Handl, M.; Corasaniti, M.; Friedrich, K.A. 2014. Insight into the structure and nanoscale conductivity of fluorinated ionomer membranes. *Journal of the Electrochemical Society,* 161:F1214–F1223.
35. Choi, P.; Jalani, N.H.; Datta, R. 2005. Thermodynamics and proton transport in Nafion II. Proton diffusion mechanisms and conductivity. *Journal of the Electrochemical Society* 152:E123–E130.
36. Thampan, T.; Malhotra, S.; Tang, H.; Datta, R. 2000. Modeling of conductive transport in proton-exchange membranes for fuel cells. *Journal of the Electrochemical Society* 147:3242.
37. Kusoglu, A.; Weber, A.Z. 2017. New insights into perfluorinated sulfonic-acid ionomers. *Chemical Reviews* 117:987–1104.
38. Yang, S.J.; Ding, X.; Han, B.H. 2018. Conjugated microporous polymers with dense sulfonic acid groups as efficient proton conductors. *Langmuir* 34:7640–7646.
39. Katzenberg, A.; Chowdhury, A.; Fang, M.; Weber, A.Z.; Okamoto, Y.; Kusoglu, A.; Modestino, M.A. 2020. Highly permeable perfluorinated sulfonic acid ionomers for improved electrochemical devices: Insights into structure–property relationships. *Journal of the American Chemical Society* 142:3742–3752.
40. Evonik wants to make green hydrogen more affordable. https://corporate.evonik.de/en/media/press-releases/corporate/evonik-wants-to-make-green-hydrogen-more-affordable-134037.html (Accessed: January 2021).
41. Zhang, K.; McDonald, M.B.; Genina, I.E.; Hammond, P.T. 2018. A highly conductive and mechanically robust OH–conducting membrane for alkaline water electrolysis. *Chemistry of Materials* 30:6420–6430.
42. Lee, M.T. 2020. Designing anion exchange membranes with enhanced hydroxide ion conductivity by mesoscale simulations. *The Journal of Physical Chemistry C* 124:4470–4482.
43. Li, L.; Wang, J.; Ma, L.; Bai, L.; Zhang, A.; Qaisrani, N.A.; Yan, X.; Zhang, F.; He, G. 2021. Dual-side-chain-grafted poly(phenylene oxide) anion exchange membranes for fuel-cell and electrodialysis applications. *ACS Sustainable Chemistry & Engineering* 9:8611–8622.
44. Long, C.; Zhao, T.; Tian, L.; Liu, Q.; Wang, F.; Wang, Z.; Zhu, H. 2021. Highly stable and conductive multicationic poly(biphenyl indole) with extender side chains for anion exchange membrane fuel cells. *ACS Applied Energy Materials* 4:6154–6165.

45. Watanabe, M.; Thomas, M.L.; Zhang, S.; Ueno, K.; Yasuda, T.; Dokko, K. 2017. Application of ionic liquids to energy storage and conversion materials and devices. *Chemical Reviews* 117:7190–7239.
46. Ito, S., Zakeeruddin, S., Comte, P.; Liska, P.; Kuang, D.; Grätzel, M. 2008. Bifacial dye-sensitized solar cells based on an ionic liquid electrolyte. *Nature Photonics* 2:693–698.
47. Feng, J.; Wang, Y.; Xu, Y.; Sun, Y.; Tang, Y.; Yan, X. 2021. Ion regulation of ionic liquid electrolytes for supercapacitors. *Energy & Environmental Science* 14:2859–2882.
48. Fedorov, M.V.; Kornyshev, A.A. 2014. Ionic liquids at electrified interfaces. *Chemical Reviews* 114:2978–3036.
49. Izgorodin, A.; Izgorodina, E.; MacFarlane, D.R. 2012. Low overpotential water oxidation to hydrogen peroxide on a MnO_x catalyst. *Energy & Environmental Science* 5:9496–9501.
50. Yan, X.; Zhang, F.; Zhang, H.; Tang, H.; Pan, M.; Fang, P. 2019. Improving oxygen reduction performance by using protic poly(ionic liquid) as proton conductors. *ACS Applied Materials and Interfaces* 11:6111–6117.
51. Zhang, R.G.; Etzold, B.J.M. 2021. Emerging applications of solid catalysts with ionic liquid layer concept in electrocatalysis. *Advanced Functional Materials* 31:2010977.
52. Amarasekara, A.S. 2016. Acidic ionic liquids. *Chemical Reviews* 116:6133–6183.
53. MacFarlane, D.R; Tachikawa, N.; Forsyth, M.; Pringle, J.M.; Howlett, P.C.; Elliott, G.D.; Davis Jr., J.H.; Watanabe, M.; Simon, P.; Angell, C.A. 2014. Energy applications of ionic liquids. *Energy & Environmental Science* 7:232–250.

CHAPTER 5

Materials for Supercapacitors

5.1 WHY SUPERCAPACITORS?

We know that many energy provision schemes require one or more chemical-to-electrical energy conversion steps or charge-discharge cycles from the previous discussions. During such stages, there are noticeable energy losses. The situation often becomes even more complicated if, for example, the stored electrical energy should be employed quickly with high current densities and power. This can happen in both automotive and stationary applications. Naturally, the minimization of losses during the energy conversion and storage steps is essential.

In principle, it is widely known that common capacitors can offer minimal energy losses between the charge-discharge cycles. Extrapolating this sort of simple idea to a larger scale, one could dream of storing electrical energy using a capacitor with a huge capacitance. In other words, why not develop supercapacitors to at least partly address the challenge of energy losses during charge and discharge cycles. In fact, such supercapacitors (SCs) do exist [1,2] and occupy more and more niches in the world energy market [3,4]. They are nowadays used in automotive, for example, public transport applications (see Figure 5.1) [5,6]. They are also used in a portfolio of devices for large-scale stationary applications when an electrical energy provision system must respond immediately with high power [7].

In terms of gravimetric power and energy density, one can place SCs between common capacitors, which can provide high specific power but

DOI: 10.1201/9781003025498-5

FIGURE 5.1 A commercial supercapacitor-powered electric bus in operation. The bus's supercapacitor requires approximately 5 minutes of charging, providing 12–17 km of driving distance. The main advantage of such a system is a potentially "unlimited" number of charging-discharging cycles within the relevant vehicle lifetime.

only for a very short time, and batteries, which can provide moderate power for extended periods (Figure 5.2). One of the main advantages of supercapacitors, if compared to batteries, is that they can be, in some cases, charged-discharged up to a few million times without a significant loss of their performance. The high cycling stability has encouraged further research and development in this field recently. Therefore, the main search direction for new electrode and electrolyte materials for SCs is finding those, which would increase the overall specific energy while maintaining high specific power. Additional advantages of supercapacitors are described in Table 5.1.

The practical use of supercapacitors for electrical charge storage was demonstrated and patented in 1957 [8]. Notably, there was no clear understanding of *how does it work* until the granting of a new patent to SOHIO in 1966 [9], in which it was acknowledged that these devices store energy in the so-called electrical double layer at the interface between the electrode and electrolyte solution. The terms "supercapacitor", and perhaps

TABLE 5.1 Advantages and Challenges of Common Supercapacitors

Advantages	**Disadvantages**
• Can provide almost unlimited cycle lifetime • Have high specific power • Relatively safe • Can efficiently operate in the temperature range from −40°C to 60°C • Can be charged in seconds; simple charging is, in many cases, possible	• Have low specific energy • High self-discharge is often observed • Provide relatively low cell voltage • Have linear charge-discharge characteristics

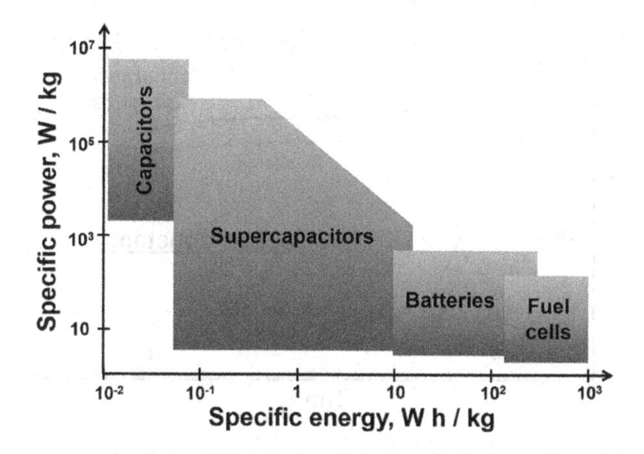

FIGURE 5.2 Supercapacitors are located in between common capacitors and batteries in terms of specific power and specific energy.

"ultracapacitor" and "electrochemical double layer capacitors" were introduced by Prof. Brian Conway. He was one of the first persons to explain *how does it work* and what role the electrical double layer and interfacial charge transfer play in the performance of these devices. Compared with conventional capacitors, the specific energy of SCs is several orders of magnitude higher (hence the 'super' or 'ultra' prefix). Supercapacitors also have higher specific power than most batteries, apart from the fact that their specific energy is lower.

One of the critical formal characteristics of supercapacitors to be distinguished from batteries is the voltage *versus* time charge-discharge characteristics at a constant current load (see Figure 5.3). As one can see from Figure 5.3, the charge-discharge profiles are drastically different. In the case of batteries, one usually distinguishes a plateau, i.e., there is a stable voltage at a constant current in the external circuit during a relatively long period of time. In contrast, typical and characteristic linear discharge profiles for SCs limit the use of all the accumulated energy in many applications. Typically, the remaining buffer charge in supercapacitors before recharging is approximately 20% of the maximum charge.

The energy stored in supercapacitors, E, and the corresponding maximum power, P_{max}, are calculated according to the well-known equations for a standard capacitance:

$$E = 0.5 \cdot C \cdot V^2$$

$$P_{max} = V^2 / 4R$$

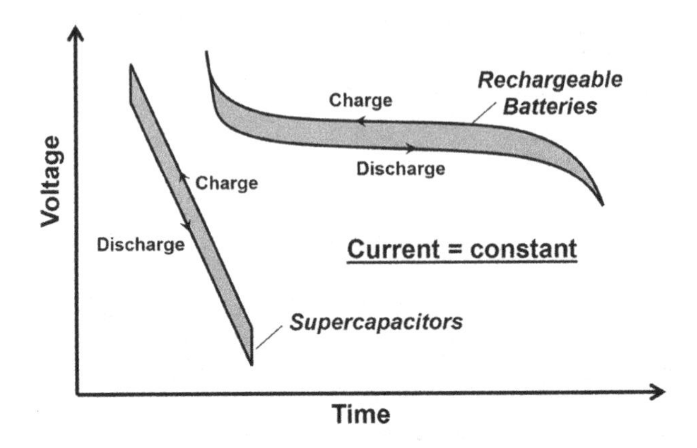

FIGURE 5.3 Schematic of voltage versus time charge-discharge characteristics for rechargeable batteries and supercapacitors.

where C is the direct current capacitance in Farads, V is the nominal voltage, and R is the equivalent series resistance in Ohms. From the equations presented above, it is clear that the operating voltage of the SCs is as essential as the high capacitance itself to achieve better device performance.

One should note that the capacitance values in the equations above are related to the direct current (DC) measurements. Why the DC-capacitance? The measured capacitance can be a function of the applied alternating current (AC) frequency. Solid electrodes (even smooth monocrystalline electrodes) can demonstrate a so-called frequency capacitance dispersion. This means that the measured capacitance values depend on the frequency used for the measurements. The origin of this phenomenon is, however, still under debate [10,11]. The situation likely originates from the fact that there are no ideal capacitances in the electrochemical systems, and the well-established description is just a first approximation in this case.

5.2 THE ELECTRICAL DOUBLE LAYER CAPACITANCE

Numerous observations that in the absence of electrode redox reactions, the electrode/electrolyte systems show a capacitive behavior could be traced back to the 19th century, well before the patents on supercapacitors were filed. At that time, researchers noticed that the relaxation "current *versus* time" curves recorded after the application of a constant external electric bias resembled those typical for standard capacitors. They most likely observed dependencies similar to the one shown in Figure 5.4, when at some point the measured current becomes zero. Moreover, suppose one integrates this kind of relaxation curve. In that case, it is possible to obtain

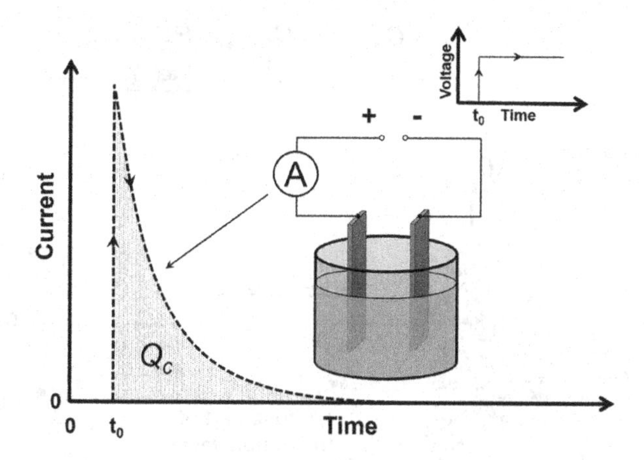

FIGURE 5.4 In the absence of electrode reactions, the relaxation current versus time curves for a two-electrode electrochemical system recorded after applying a constant bias resembles the charging curves typical for standard capacitors. Integration of the charging curve also gives the charge, Q_c, necessary to calculate the effective capacitance.

the charge, Q_c, associated with such a system relaxation and subsequently calculate the effective capacitance of this system. However, it was probably remarkable to observe that the effective capacitance did not really depend on the distance between two electrodes. That means one could not draw direct correlations to classical capacitors.

Considering these observations, scientists tried to develop a theory explaining such behavior of electrochemical systems. It took decades until a reasonable idea had been elaborated after such observations. It started from the models of the electrochemical double layer of Helmholtz (1853) [12], and only after 1924, Stern [13] suggested a theory describing the observed phenomena fairly well. The theory of Stern was the first to be confirmed using various systems [14] and model electrodes by 1954 [15].

According to the current understanding, based on the same ideas, the electrified interface between an electron conductor and liquid electrolyte can be schematically represented by a model described in Figure 5.5.

In the simplest case, the double electric layer at the interface can be imagined as a layer of solvent molecules with some ions, which are dynamically structured. Historically, the scientific community distinguishes the inner Helmholtz layer, which can be quantified if one draws an arbitrary "plane" behind the first layer of adsorbed solvent molecules, as schematically shown in Figure 5.5. Its thickness can be assessed to be in the range of several angstroms (10^{-10} m). One can also define the outer Helmholtz

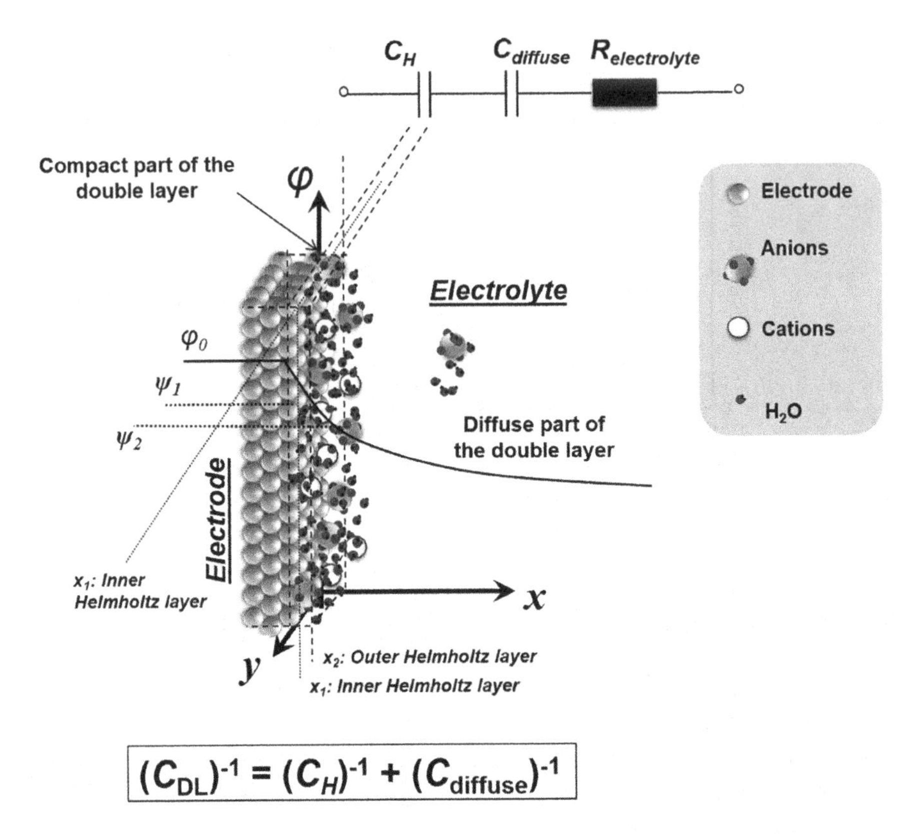

FIGURE 5.5 If there are no electrode redox reactions, the electrified interface between an electron-conducting electrode and an ionically conducting electrolyte can be schematically described by this picture. The equivalent electric circuit describes the system well, but there are no ideal capacitances in it. C_H, $C_{diffuse}$, and C_{DL} designate the effective Helmholtz layers' capacitances, the diffuse part of the double layer, and the resulting double layer capacitances.

layer if one distinguishes the virtual plane passing through the centers of ions located in the closest proximity to the electrode surface. The thickness of this layer is usually less than 1 nm. Beyond this, one can consider the so-called diffuse part of the double layer, where the concentration of the charged species approaches the bulk concentration of the electrolyte solutions (Figure 5.5).

Taking a closer look at the model described in Figure 5.5, one can conclude that similarities with classical capacitors are not straightforward. However, the capacitance approximations are practically useful. While the interface behavior only resembles [16] that of the standard capacitors with an effective capacitance, C, these electrochemical interfaces can be reasonably well described by equations applicable in the case of a classical capacitor. One can use a classical equation as a roadmap in designing better materials for these devices:

$$C = \varepsilon_r \varepsilon_0 A/d$$

where the parameter d should be considered as the effective thickness of the *double layer capacitor*. As the double layer thickness is in the order of some monolayers (i.e., at the nanometer-scale), the resulting capacitance of SCs is very high. The parameter A in the formula is the electrodes' active surface area, ε_0 is the dielectric permittivity of vacuum and ε_r is the relative permittivity, which mainly depends on the solvent nature and electrolyte composition.

Let us now consider a simple scheme of the double-layer supercapacitor, as described in Figure 5.6. It consists of a pair of identical electrodes in contact with an electrolyte. Additionally, there is an electronically insulating separator [17], which is permeable for the ions. Its role is to prevent a sudden contact between the electrodes during the use of the device.

At the interfaces between the electrodes and electrolyte, the electrical double layers will be formed. In aqueous solutions and in the case of relatively inert anions and cations, the latter are normally solvated by water molecules, as schematically illustrated in Figure 5.6. This solvation shell will dynamically isolate the ions from direct contact with the electrode material, defining the thickness of the inner Helmholtz layer.

Figure 5.6a schematically shows the situation when such SC is not charged. In such a case, one can assume that the distribution of the anions and cations between the electrodes should be symmetrical. The equivalent electric circuit of such a system is given below in Figure 5.6a. It consists of two identical double-layer capacitances and the resistance of the electrolyte. By applying an external bias, it is evident to change the net situation. Anions will accumulate at the positive terminal, and cations will tend to move to the negative one. While this kind of movement is well expectable, it is also equivalent to the charging of the virtual double layer capacitances, C_{DL1} and C_{DL2}. As discussed before, an external observer will record a typical charging curve of a capacitor with very high capacitance.

What kind of electrode energy materials is then promising to use for supercapacitors? In double-layer capacitors, it is obvious to use light electronic conductors with a high surface area. The current state-of-the-art electrodes typically involve carbon nanomaterials. Common design principles

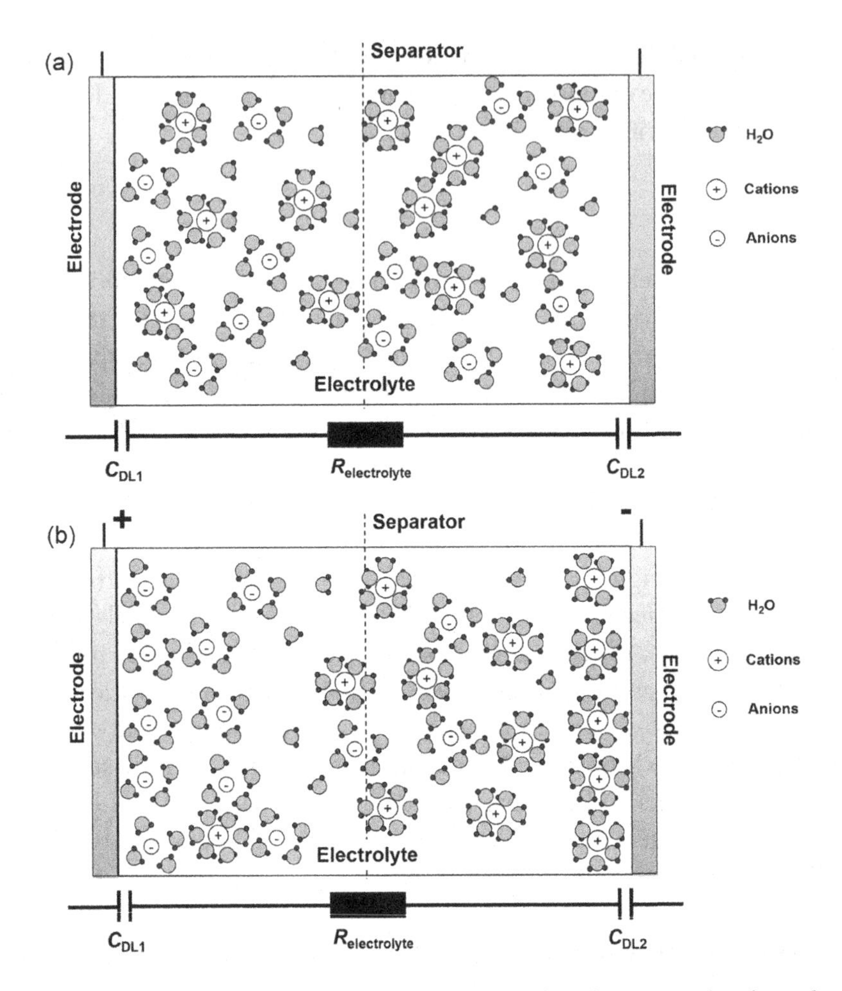

FIGURE 5.6 A schematic representation of ion redistribution in the electrolyte when the double layer supercapacitor is (a) discharged and (b) fully charged. Equivalent electric circuits are presented below of each model. C_{DL1} and C_{DL2} designate the double layer capacitances of the electrodes.

tend to create structures similar to that schematically shown in Figure 5.7 [18]. For example, carbon nanotubes or fibers [19] supported on a current collector can provide a relatively high capacitance due to the increased surface area.

According to the equation above, defining the classical capacitance, one approaches a limit in designing the materials for the double layer SCs. If not increasing the surface area of the electrodes, what can one envisage to increase the overall effective capacitance of the devices further? One cannot significantly reduce the parameter d, which is related to electrolytes

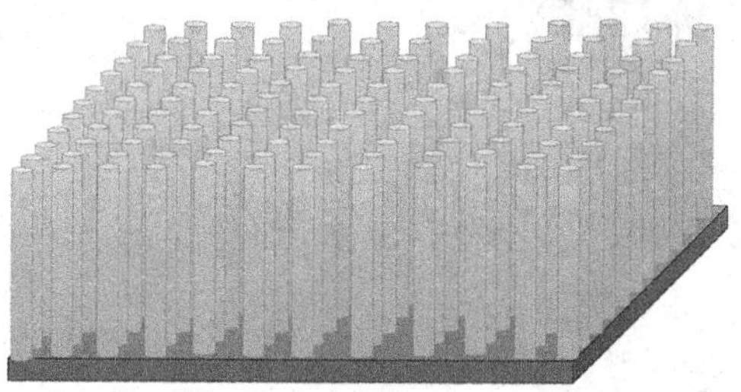

FIGURE 5.7 Schematic illustrates one of the straightforward approaches to increase the surface area of the electrodes in the double-layer supercapacitors and hence increase the SC device's overall capacitance. Carbon nanotubes are suitable for such purposes.

and already in the nanometers range. The relative permittivity depends on the nature of the electrolyte and provides only a limited degree of freedom to design SCs with much better performance. However, there is another option, which is less obvious and based on the so-called *pseudocapacitive* behavior of some electrochemical systems.

5.3 PSEUDOCAPACITANCES

The idea to use pseudocapacitive effects to increase the capacitances of SCs again probably belongs to Prof. Conway, who suggested using the so-called fast *surface limited* Faradaic reactions. The approach is relatively simple: why not use some fast electrode reactions, which can occur either at the surface, in the double layer, or in thin layers of some electroactive oxide or conducting polymer (for example, polyaniline, polypyrrole, or polythiophene [20]) materials when external bias is applied [21]. To start explaining the phenomenon, one can take a look at Figure 5.8.

An interfacial charge transfer can occur when an ion penetrates the surface water layer and is directly in contact with the electrode surface. It isn't easy to say whether such *specifically adsorbed* species belong to either electrode or the electrolyte side in many cases. However, *specific adsorption* causes an additional external current to be observed added to the double layer charge-discharge current. There is a limited amount of adsorption places at the surface. Therefore, this additional external current due to specific adsorption will go down to zero quickly. If one records the electric current in the external circuit, a curve shown in Figure 5.4 with the increased Q_c value is detected, indicating the enlarged effective capacitance.

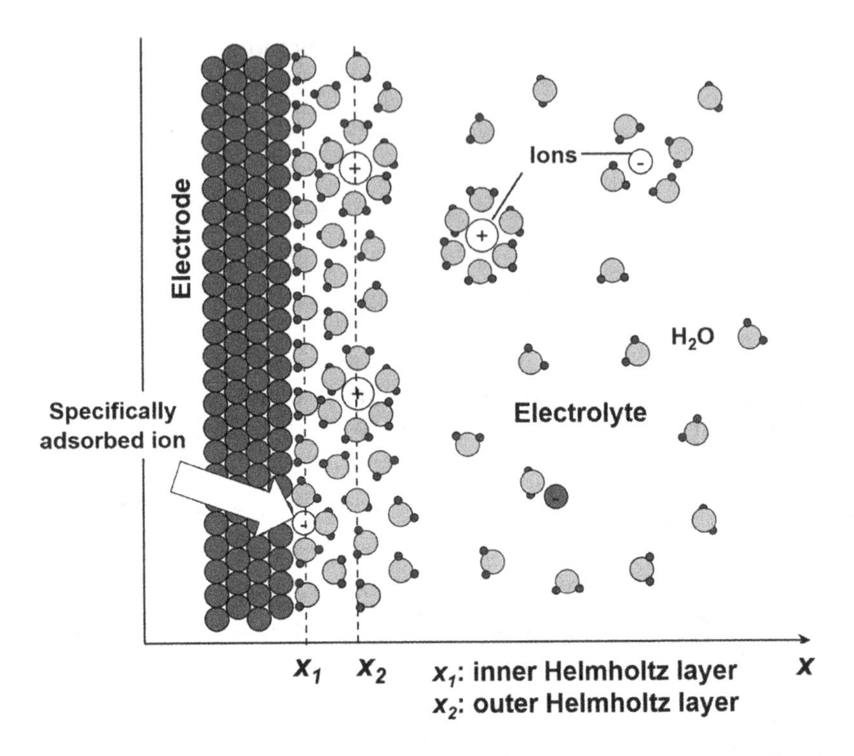

FIGURE 5.8 Schematic representation of a specifically adsorbed ion at the electrode surface. The specifically adsorbed ion goes to the first water layer. In this case, the thickness of the inner Helmholtz layer is conventionally defined using a line going through the center of the ion, as shown in the figure.

Specific adsorption or other surface limited Faradaic reactions can all be considered as a "leakage" of the double layer capacitance. If one organizes this leakage controllably and considering that this kind of reaction will be fast and relatively short in time, one can envisage even bigger effective capacitances. The current-voltage characteristics for such surface-limited reactions must be close to that shown in Figure 5.9a when just a small change in the potential causes an immediate redox response.

Let us consider a model situation shown in Figure 5.10. The figure describes a system with the electrodes coated with thin films of electroactive material, in which fast redox transformations occur upon even slight electrode potential changes. The system also exhibits the specific adsorption of cations and anions. The Faradaic processes in the films and at the surface can be considered as the double layer capacitance leakage. Therefore, parallel branches in the equivalent circuit appear. They consist

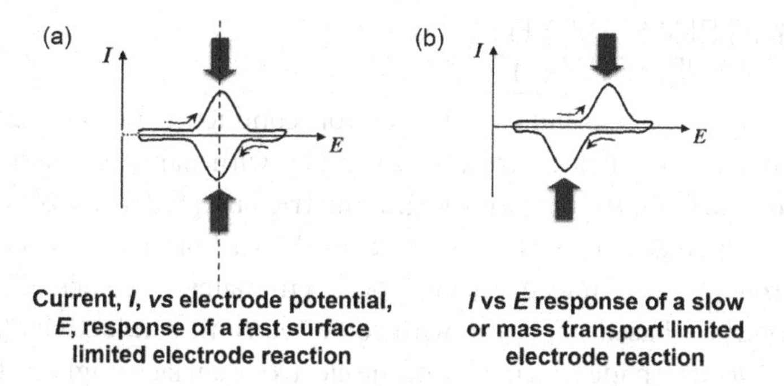

FIGURE 5.9 Basic voltammetric characteristics of (a) ideal and (b) not optimal surface limited Faradaic processes to increase the capacitance of the SCs.

of adsorption pseudocapacitances and pseudoresistances. The pseudoresistances reflect the fact that the redox transformations at the interface and in the film are still somewhat slower than the dipole rearrangements in the double layer. The pseudocapacitances are used to account for the excessive charge passed in parallel to the double layer charging current due to redox transformations in the film or related to the specific adsorption.

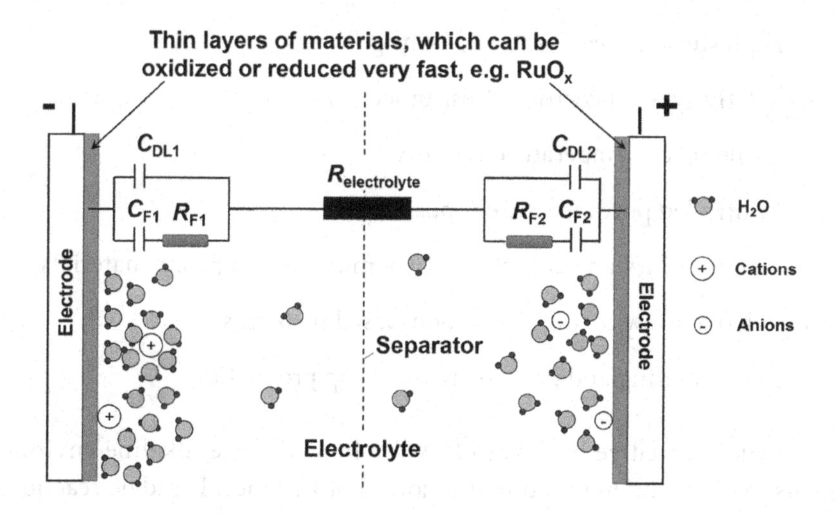

FIGURE 5.10 The use of the so-called surface limited reactions, when electroactive materials can be reduced and oxidized very fast, can increase the effective capacitance of the system. C_{F1} and C_{F2} are the "pseudocapacitances" and $R_{F1,}$ and R_{F2} are "pseudoresistances" accounting for the fast Faradaic processes in and at the thin electroactive film on the surface of the electrodes.

5.4 MATERIALS AND ELECTROLYTES FOR SUPERCAPACITORS

Apparently, one needs lightweight electron-conducting porous [22] materials with a large surface area, for example, with nano- or mesoporous structures [23,24], to increase the gravimetric energy density of the double layer supercapacitors. It is easy to consider carbon-based materials for such a role [25–28]. Indeed, recent progress in material science offers very porous carbon-based structures with good electronic conductivity [29–31]. In other words, an ideal structure of the electrode surface might be the one schematically shown in Figure 5.7. This kind of design largely increases the effective surface area and thereby the overall effective capacitance [32]. One can hardly imagine some affordable electronically-conducting material with a high surface area other than carbon nanotubes. Another option is to use graphene-based structures [33–38].

The attraction of carbon materials as supercapacitor electrode materials also arises from a unique combination of chemical and physical properties, some of which are:

- High electronic conductivity
- High surface area ($\sim$1 to > 2,000 m^2g^{-1})
- Relatively good corrosion resistance
- Rather good temperature stability
- Controlled pore structure is possible
- Processability and compatibility of multiple composite materials
- Relatively low cost of the carbon-based materials
- Easy handling and processing for cheap production

The pseudocapacitive behavior of oxides in acidic and alkaline environments involves different redox reactions. For instance, Faradaic reactions taking place in the system involving ruthenium oxides and oxyhydroxides, one of the state-of-the-art materials, can be described as follows (this transformation is believed to be almost as fast as the rearrangements of dipoles in the electrical double layer):

$$RuO_2 + xH^+ + xe^- \leftrightarrow RuO_{2-x}(OH)_x$$

There are less expensive alternatives, which involve slower reactions in the oxyhydroxide and oxide thin films of cobalt, nickel [39], and manganese. One example of the redox transformations in the case of cobalt-based thin films used in SCs is shown below:

$$Co(OH)_2 + OH^- \leftrightarrow CoOOH + H_2O + e^-$$

$$CoOOH + OH^- \leftrightarrow CoO_2 + H_2O + e^-$$

$$Co_3O_4 + H_2O + OH^- \leftrightarrow 3CoOOH + e^-$$

An approach to benefit from both pseudocapacitance and the large surface area is to use carbon nanomaterials with oxides, electroactive polymers or metal-organic frameworks and their derivatives [40–42] deposited on top.

What about the electrolyte choice? Of course, aqueous electrolytes provide high ionic conductivity and safety [43,44]. They are also affordable, reducing the final price of the SC devices. Aqueous electrolytes enable very good kinetic behavior of the electrolyte ions leading to high charge-discharge rates. This behavior is due to the relatively high conductivity and low viscosity of the concentrated solutions. For example, the conductivity of 1M H_2SO_4 is ~730 mS cm^{-1} compared to the much lower typical values of 10–20 mS cm^{-1} for organic solutions. However, a relatively narrow stability window of aqueous electrolytes limits their application since the power of the devices scales with the operating voltage as V^2, as explained above. For aqueous electrolytes, the maximum operating voltage is theoretically limited by water decomposition (in practice, *ca* 2.0 V, including the overpotential for the water splitting).

In contrast, many liquid organic electrolytes and solid polymer electrolytes provide a much wider stability window [45,46]. The use of nonaqueous electrolytes in supercapacitors has the main advantage of higher operating voltages compared to aqueous systems [47]. Voltage windows can range up to few volts, and since also the stored energy increases as V^2, it is possible to attain quite large energy and power densities. The major disadvantages of nonaqueous systems are their lower conductivity, higher viscosity resulting in increased equivalent series resistance, reduced wettability, and flammability. A decrease in wettability will effectively reduce the electrode surface area in contact with the electrolyte, decreasing the energy and power density. Most commercial systems that use organic electrolytes are manufactured in inert atmospheres and are costly to be produced.

Other alternatives to be used as electrolytes in SCs are ionic liquids [48–50] or sometimes blends of ionic liquids and organic solvents [51]. The main advantages of ionic liquids are the good solvating properties, relatively high conductivity, nonvolatility, low toxicity, large potential window, negligible vapor pressure, and good electrochemical stability [52]. Capacitances approaching 100 F/g for activated carbon (AC) electrodes have been reported, with a potential window that is much greater than that of aqueous systems. Disadvantages of ionic liquids in SCs include high viscosities and low conductivities compared to aqueous electrolytes.

5.5 SUMMARY AND CONCLUSIONS

Supercapacitors are an attractive class of energy storage devices, which can quickly provide high power and high current densities, with the capacitances being significantly higher compared to common, e.g., ceramic capacitors. An essential advantage of supercapacitors is the possibility of charging and discharging up to a few million times without significant losses in their performance. However, further substantial research and development activities are required for their broader commercialization. Nevertheless, they are in commercial use in buses, cars, and other automotive applications. Carbon-based materials are among the state-of-the-art ones for the electrodes exclusively utilizing the capacitive effects in the double electric layer. It is possible to increase the effective capacitance using pseudocapacitive effects due to surface limited Faradaic reactions. RuO_2 is the best-known material if pseudocapacitive effects are used. However, Co-, Ni-, Mn- oxide materials and other oxide materials are used along with conducting polymers immobilized at the electrode surface. Promising areas of applications of SCs also include stationary energy storage applications, when an immediate response of the system electricity grid with high power and high current densities is necessary to compensate for sudden fluctuations, as they increasingly appear in recent years due to the integration of renewable energy sources.

5.6 QUESTIONS

1. What are the working principles of supercapacitors?
2. How to model the electric double layer between solid electron conductors and aqueous electrolytes?
3. How to increase the effective double layer capacitance?

4. What are the common requirements for the electrode materials used in supercapacitors?
5. What are the surface limited Faradaic processes? How to use them to increase the performance of supercapacitors?
6. Name state-of-the-art electrode materials for supercapacitors.
7. What are typical electrolytes for supercapacitors?
8. Analyze what limitations and possible application areas of supercapacitors are?

REFERENCES

1. Conway, B. 1999. *Electrochemical Supercapacitors: Scientific Fundamentals and Technological Applications (POD)*. Kluwer Academic/Plenum: New York.
2. Burt, R.; Birkett, G.; Zhao, X.S. 2014. A review of molecular modelling of electric double layer capacitors. *Physical Chemistry Chemical Physics* 16:6519–6538.
3. Zhao, J.; Burke, A.F. 2021. Review on supercapacitors: Technologies and performance evaluation. *Journal of Energy Chemistry* 59:76–291.
4. Wang, F.; Wu, X.; Yuan, X.; Liu, Z.; Zhang, Y.; Fu, L.; Zhu, Y.; Zhou, Q.; Wu, Y.; Wei, H. 2017. Latest advances in supercapacitors: From new electrode materials to novel device designs. *Chemical Society Reviews* 46: 6816–6854.
5. Wang, Y.; Song, Y.; Xia, Y. 2016. Electrochemical capacitors: Mechanism, materials, systems, characterization and applications. *Chemical Society Reviews* 45:5925–5950.
6. Zhang, L.; Hu, X.; Wang, Z.; Sun, F.; Dorrell, D.G. 2018. A review of supercapacitor modeling, estimation, and applications: A control/management perspective. *Renewable and Sustainable Energy Reviews* 81:1868–1878.
7. Libich, J.; Máca, J.; Vondrák, J.; Čech, O.; Sedlaříková, M. 2018. Supercapacitors: Properties and applications. *Journal of Energy Storage* 17:224–227.
8. Becker, H.I. 1957. Low voltage electrolytic capacitor. *United States Patent* 2,800,616.
9. Rightmire, R.A. 1966. Electrical energy storage apparatus. *United States Patent* 3,288,641.
10. Bandarenka, A.S. 2013. Exploring the interfaces between metal electrodes and aqueous electrolytes with electrochemical impedance spectroscopy. *Analyst* 138:5540–5554.
11. Pajkossy, T.; Jurczakowski, R. 2017. Electrochemical impedance spectroscopy in interfacial studies. *Current Opinion in Electrochemistry* 1:53–58.
12. Helmholtz, H. 1853. Ueber einige gesetze der vertheilung elektrischer ströme in körperlichen leitern mit anwendung auf die thierisch-elektrischen versuche. *Annalen der Physik und Chemie* 165:211–233.

13. Stern, O. 1924. Zur Theorie der elektrolytischen Doppelschicht. *Zeitschrift für Elektrochemie und Angewandte Physikalische Chemie* 30:508–516.
14. Grahame, D.C. 1947. The electrical double layer and the theory of electrocapillarity. *Chemical Reviews* 41:441–501.
15. Grahame, D.C. 1954. Differential capacity of mercury in aqueous sodium fluoride solutions. I. Effect of concentration at 25°. *Journal of the American Chemical Society* 76:4819–4823.
16. Cl, M.; Péan, C.; Rotenberg, B.; Madden, P.A.; Simon, P.; Salanne, M. 2013. Simulating supercapacitors: Can we model electrodes as constant charge surfaces? *The Journal of Physical Chemistry Letters* 4:264–268.
17. Li, L.; Lu, F.; Wang, C.; Zhang, F.; Liang, W.; Kuga, S.; Dong, Z.; Zhao, Y.; Huang, Y.; Wu, M. 2018. Flexible double-cross-linked cellulose-based hydrogel and aerogel membrane for supercapacitor separator. *Journal of Materials Chemistry A* 6:24468–24478.
18. Zhou, G.; Xu, L.; Hu, G.; Mai, L.; Cui, Y. 2019. Nanowires for electrochemical energy storage. *Chemical Reviews* 119:11042–11109.
19. Chen, S.; Qiu, L.; Cheng, H.M. 2020. Carbon-based fibers for advanced electrochemical energy storage devices. *Chemical Reviews* 120: 2811–2878.
20. Meng, Q.; Cai, K.; Chen, Y.; Chen, L. 2017. Research progress on conducting polymer based supercapacitor electrode materials. *Nano Energy* 36:268–285.
21. González, A.; Goikolea, E.; Barrena, J.A.; Mysyk, R. 2016. Review on supercapacitors: Technologies and materials. *Renewable and Sustainable Energy Reviews* 58:1189–1206.
22. Shao, H.; Wu, Y.C.; Lin, Z.; Taberna, P.L. Simon, P. 2020. Nanoporous carbon for electrochemical capacitive energy storage. *Chemical Society Reviews* 49:3005–3039.
23. Fang, Y.; Zhang, Q.; Cui, L. 2021. Recent progress of mesoporous materials for high performance supercapacitors. *Microporous and Mesoporous Materials* 314:110870.
24. Yu, Z.; Tetard, L.; Zhai, L.; Thomas, J. 2015. Supercapacitor electrode materials: Nanostructures from 0 to 3 dimensions. *Energy and Environmental Science* 8:702–730.
25. Béguin, F.; Presser, V.; Balducci, A.; Frackowiak, E. 2014. Carbons and electrolytes for advanced supercapacitors. *Advanced Materials* 26:2219–2251.
26. Péan, C.; Merlet, C.; Rotenberg, B.; Madden, P.; Taberna, P.-L.; Daffos, B.; Salanne, M.; Simon, P. 2014. On the dynamics of charging in nanoporous carbon-based supercapacitors. *ACS Nano* 8:1576–1583.
27. Liu, L.; Niu, Z.; Chen, J. 2016. Unconventional supercapacitors from nanocarbon-based electrode materials to device configurations. *Chemical Society Reviews* 45:4340–4363.
28. Shao, Y.; El-Kady, M.F.; Wang, L.J.; Zhang, Q.; Li, Y.; Wang, H.; Mousavi, M.F.; Kaner, R.B. 2015. Graphene-based materials for flexible supercapacitors. *Chemical Society Reviews* 44:3639–3665.

29. Yang, Z.; Ren, J.; Zhang, Z.; Chen, X.; Guan, G.; Qiu, L.; Zhang, Y.; Peng, H. 2015. Recent advancement of nanostructured carbon for energy applications. *Chemical Reviews* 115:5159–5223.
30. Zhang, L.L.; Zhao, X.S. 2009. Carbon-based materials as supercapacitor electrodes. *Chemical Society Reviews* 38:2520–2531.
31. Peng, X.; Peng, L.; Wu, C.; Xie, Y. 2014. Two dimensional nanomaterials for flexible supercapacitors. *Chemical Society Reviews* 43:3303–3323.
32. Chabi, S.; Peng, C.; Hu, D.; Zhu, Y. 2014. Ideal three-dimensional electrode structures for electrochemical energy storage. *Advanced Materials* 26:2440–2445.
33. Fileti, E.E. 2020. Electric double layer formation and storing energy processes on graphene-based supercapacitors from electrical and thermodynamic perspectives. *Journal of Molecular Modeling* 26:159.
34. Zhang, C.; Peng, Z.; Lin, J.; Zhu, Y.; Ruan, G.; Hwang, C-C.; Lu, W.; Hauge, R.H.; Tour, J.M. 2013. Splitting of a vertical multiwalled carbon nanotube carpet to a graphene nanoribbon carpet and its use in supercapacitors. *ACS Nano* 7:5151–5159.
35. Sun, Z.; Fang, S.; Hu, Y.H. 2020. 3D graphene materials: From understanding to design and synthesis control. *Chemical Reviews* 120:10336–10453.
36. Chen, K.; Song, S.; Liu, F.; Xue, D. 2015. Structural design of graphene for use in electrochemical energy storage devices. *Chemical Society Reviews* 44:6230–6257.
37. Ke, Q.; Wang, J. 2016. Graphene-based materials for supercapacitor electrodes – a review. *Journal of Materiomics* 2:37–54.
38. Olabi, A.G.; Abdelkareem, M.A.; Wilberforce, T.; Sayed, E.T. 2021. Application of graphene in energy storage device – A review. *Renewable and Sustainable Energy Reviews* 135:110026.
39. Yi, T.F.; Wei, T.T.; Mei, J.; Zhang, W.; Zhu, Y.; Liu, Y.G.; Luo, S.; Liu, H.; Lu, Y.; Guo, Z. 2020. Approaching high-performance supercapacitors via enhancing pseudocapacitive nickel oxide-based materials. *Advanced Sustainable Systems* 4:1900137.
40. Choudhary, R.B.; Ansari, S.; Purty, B. 2020. Robust electrochemical performance of polypyrrole (PPy) and polyindole (PIn) based hybrid electrode materials for supercapacitor application: A review. *Journal of Energy Storage* 29:101302.
41. De Adhikari, A.; Morag, A.; Seo, J.; Kim, J.M.; Jelinek, R. 2020. Polydiacetylene–Perylenediimide Supercapacitors. *ChemSusChem* 13:3230–3236.
42. Choudhary, R.B.; Ansari, S.; Majumder, M. 2021. Recent advances on redox active composites of metal-organic framework and conducting polymers as pseudocapacitor electrode material. *Renewable and Sustainable Energy Reviews* 145:110854.
43. Pal, B.; Yang, S.; Ramesh, S.; Thangadurai, V.; Jose, R. 2019. Electrolyte selection for supercapacitive devices: A critical review. *Nanoscale Advances* 1:3807–3835.

44. Park, J.; Lee, J.; Kim, W. 2021. Water-in-salt electrolyte enables ultrafast supercapacitors for AC line filtering. *ACS Energy Letters* 6:769–777.
45. Xia, L.; Yu, L.; Hua, D.; Chen, G.Z. 2017. Electrolytes for electrochemical energy storage. *Materials Chemistry Frontiers* 1:584–618.
46. Sato, T.; Marukane, S.; Morinaga, T.; Kamijo, T.; Arafune, H.; Tsujii, Y. 2015. High voltage electric double layer capacitor using a novel solid-state polymer electrolyte. *Journal of Power Sources* 295:108–116.
47. Lin, Z.; Goikolea, E.; Balducci, A.; Naoi, K.; Taberna, P.L.; Salanne, M.; Yushin, G.; Simon, P. 2018. Materials for supercapacitors: When Li-ion battery power is not enough. *Materials Today* 21:419–436.
48. Salanne, M. 2017. Ionic liquids for supercapacitor applications. *Topics in Current Chemistry* 375:63.
49. Krause, A.; Balducci, A. 2011. High voltage electrochemical double layer capacitor containing mixtures of ionic liquids and organic carbonate as electrolytes. *Electrochemistry Communications* 13:814–817.
50. Stettner, T.; Balducci, A. 2021. Protic ionic liquids in energy storage devices: Past, present and future perspective. *Energy Storage Materials* 40:402–414.
51. Schütter, C.; Neale, A.R.; Wilde, P.; Goodrich, P.; Hardacre, C.; Passerini, S.; Jacquemin, J.; Balducci, A. 2016. The use of binary mixtures of 1-butyl-1-methylpyrrolidinium bis{(trifluoromethyl)sulfonyl}imide and aliphatic nitrile solvents as electrolyte for supercapacitors. *Electrochimica Acta* 220:146–155.
52. Watanabe, M.; Thomas, M.L.; Zhang, S.; Ueno, K.; Yasuda, T.; Dokko, K. 2017. Application of ionic liquids to energy storage and conversion materials and devices. *Chemical Reviews* 117:7190–7239.

CHAPTER 6

Functional Materials for Primary and Rechargeable Batteries

6.1 BATTERIES

First, let us define what rechargeable (secondary) and nonrechargeable (primary) batteries are in order to start the discussion related to this topic.

DEFINITION:

A primary cell or battery cannot be (or is not designed to be) easily recharged after a single use and is discarded after the discharge. A secondary cell or battery can be electrically recharged after the usage to its original predischarged condition by passing a current through the circuit in the opposite direction to the current during discharge.

In the beginning, one should mention that the design principles and material choice will be, in the case of batteries, very much dependent on the field of their use [1,2]. One can distinguish several dominating areas, where different types of primary and rechargeable batteries are nowadays applied, and where the research and development are most dynamic.

- Development of better battery materials for portable applications
- New efficient battery materials for automotive applications
- Battery materials for larger-scale, stationary energy storage

DOI: 10.1201/9781003025498-6

The fundamental working principles of batteries are analogous to the previously considered energy conversion and storage devices. They are based on the usage of spatially separated spontaneous redox reactions *via* electrode/electrolyte systems. In historically "classical" batteries, typically, both the product and the reactant stay at the surface of the electrodes (Figure 6.1). It should be noted here that conventionally, the designation of electrodes to be called anodes or cathodes is related to the *discharge* situation. This means the negative terminal, where oxidation of the electroactive material occurs, is called the anode. Correspondingly, the positive terminal is designated as the cathode (see Figure 6.1).

A typical characteristic charge-discharge curve for batteries is shown in Figure 6.2. As it was mentioned in the previous chapter, even though the architecture of batteries sometimes resembles the design of certain supercapacitors, there is an essential difference between these devices. Batteries are typically designed to maintain stable voltage for a relatively long operation time, with a stable voltage plateau, as schematically shown

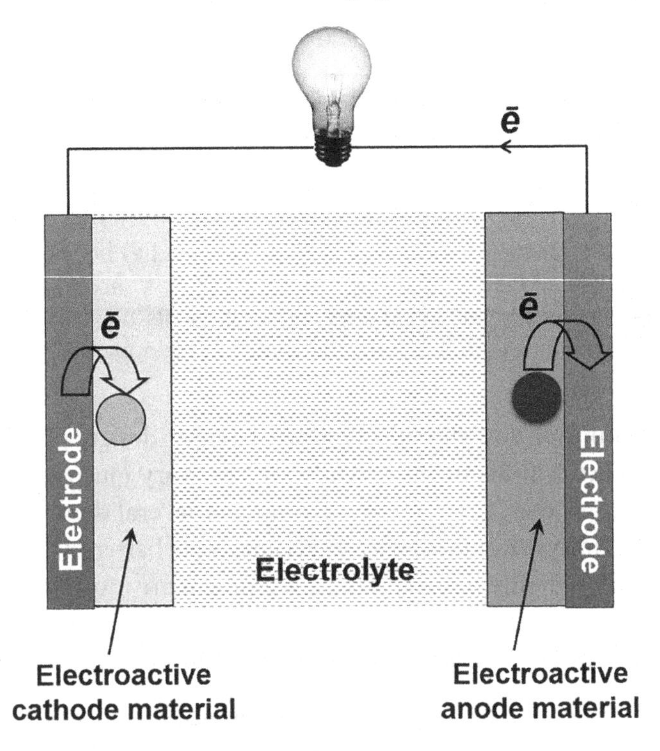

FIGURE 6.1 A simple scheme of a "classical" battery system during the discharge. Batteries use spatially separated spontaneous electrode reactions. The reactants and products, in this case, remain at the surface of the electrodes.

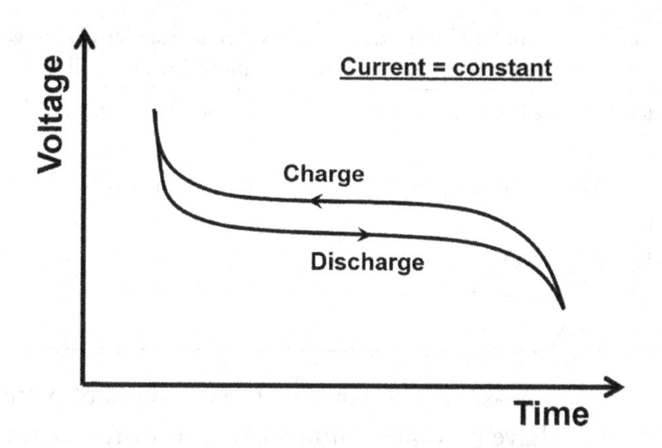

FIGURE 6.2 A simplified schematic curve, which characterizes the charging and discharging processes for a conventional rechargeable battery.

in Figure 6.2. The energy losses during the charge and discharge cycles are usually more prominent in batteries if compared to the supercapacitors. Finally, the number of possible charge-discharge cycles is much smaller in the case of batteries, practically limited to a few hundred cycles with the current technologies.

One can also admit that the working principles of batteries are also conceptually pretty similar to those of fuel cells, where the redox processes are spatially separated. While in fuel cells one can supply the reactants continuously, most of the classical batteries operate with a limited amount of electroactive species stored at the electrode surface. A mix of both concepts is used in the so-called redox-flow batteries [3,4], which will, however, not be considered here.

Remarkably, one can design a virtually unlimited number of battery types. In order to create a battery, one needs to select two electrode half-reactions with a suitable (or maximal) difference in their redox potentials. Then it is essential to carefully select electrolyte composition, current collectors, and ion-permeable but electronically insulating separators or salt bridges. Let us consider, for example, some systems containing silver compounds. Table 6.1 shows that even using redox transformations involving just silver, one can, in principle, construct a battery that would give almost 1 V as the open-circuit voltage. For that, one should, for example, use $Ag^+ + e^- \leftrightarrow Ag$ (standard potential 0.7996 V) and $AgI + e^- \leftrightarrow Ag + I^-$ (standard potential −0.152 V) half-reactions. Under standard conditions, the open-circuit voltage for an ideally designed system should be equal to $0.7996\ V - (-0.152\ V) = 0.9516\ V$.

TABLE 6.1 Standard Electrode Potentials (versus Standard Hydrogen Electrode, SHE) for Some Redox Transformations Involving Silver

Redox Electrode Reaction	Standard Potential, V (vs. SHE)
$Ag^{+}+e^{-} \leftrightarrow Ag$	0.7996
$Ag(acetate)+e^{-} \leftrightarrow Ag+(acetate)^{-}$	0.643
$AgCl+e^{-} \leftrightarrow Ag+Cl^{-}$	0.222
$AgBr+e^{-} \leftrightarrow Ag+Br^{-}$	0.0713
$AgCN+e^{-} \leftrightarrow Ag+CN^{-}$	−0.017
$AgI+e^{-} \leftrightarrow Ag+I^{-}$	−0.152

While there is a possibility to construct many types of batteries, relatively few of them have become commercially successful. Several parameters largely influence the area of application and the success of the particular technology. These are material availability and its cost, gravimetric and volumetric energy densities, the toxicity of the materials, and safety issues, to name a few.

Figure 6.3 compares several types of common rechargeable and non-rechargeable commercial batteries. As one can see, at least two types of primary batteries outperform state-of-the-art Li-ion or lead-acid rechargeable batteries in terms of gravimetric energy density. Those are lithium and alkaline manganese dioxide primary batteries. As for the latter, the MnO_2 batteries occupy a significant niche in the market nowadays due to high specific

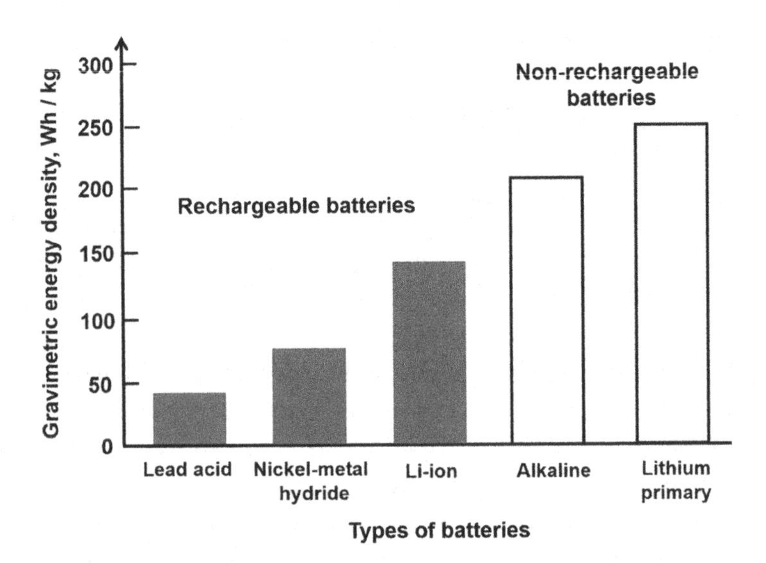

FIGURE 6.3 Specific energy comparison of some secondary and primary batteries.

energy, inexpensive materials used in them, and environmental friendliness. In the following, before further considerations on common secondary batteries, this type of nonrechargeable devices is considered in more detail.

6.2 NONRECHARGEABLE ALKALINE MANGANESE DIOXIDE BATTERY

Alkaline manganese dioxide batteries (AMDBs) are probably the most commercially successful primary batteries till date. They were first patented in 1957 [5], and the early designs involved some toxic materials such as Hg. Nowadays, such harmful additives usually are not used, and therefore, these batteries can be considered environmentally friendly.

The main advantages of the AMDB can be summarized as follows.

- They have low internal resistance, usually less than 1 Ohm
- They have a high energy density
- Have a wide working temperature range (–20°C to 54°C)
- Demonstrate slow self-discharge (96% energy retention after 1 year, long possible storage time)
- They are now rather cost-effective and relatively safe

Figure 6.4 explains the working principles of AMDBs. The battery anode typically consists of a gelled mixture of zinc powder and potassium hydroxide electrolyte. The gelling agent is usually carboxymethyl cellulose, and the anode material is in contact with a current collector generally made of brass. The cathode consists of a mixture of manganese dioxide (MnO_2) and carbon powders. Carbon (graphite) is used in the cathode because manganese dioxide is not a good electronic conductor on its own. The material of the cathode is in contact with a stainless steel current collector. Finally, the separator is ionically conducting but electronically insulating while at the same time preventing mixing of the reagents.

During the spontaneous oxidation of the zinc anode, the following electrode reaction takes place in alkaline media:

$$Zn + 2OH^- - 2e^- \rightarrow ZnO\,(ZnO \text{ at the surface passivates anode}) + H_2O$$

One should note that the rate at which the formation of the insulating zinc oxide occurs at the surface of the metal particles largely determines the

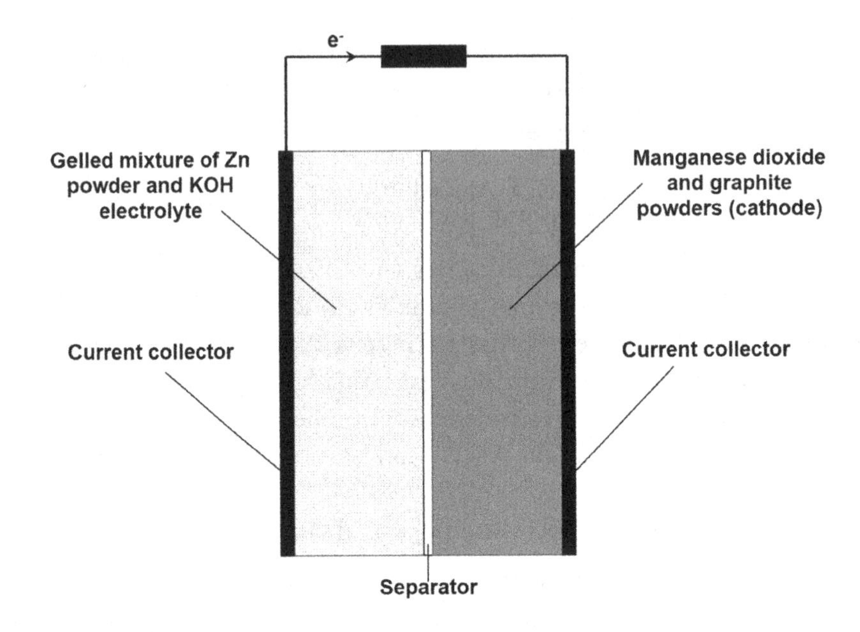

FIGURE 6.4 A general scheme of an AMDB.

resulting battery capacity. The slower the discharge, the higher the capacity of the battery.

Simultaneously, manganese dioxide is reduced, forming a nonstoichiometric compound with a tentative formula Mn_2O_3, according to the following scheme.

$$2MnO_2 + H_2O + 2e^- \rightarrow Mn_2O_3(\text{tentatively}) + 2OH^-$$

In general, the overall reaction is $Zn + 2MnO_2 \rightarrow ZnO + Mn_2O_3$, giving an initial open-circuit voltage of approximately 1.59 V.

While the electrode materials for this type of battery are cheap, affordable, and environmentally friendly, the main disadvantage, limiting the use of AMDBs to some portable applications, is that they are, of course, primary, nonrechargeable batteries. Now, let us consider what kind of rechargeable batteries are the most popular nowadays for energy storage.

6.3 RECHARGEABLE LEAD-ACID BATTERIES

Perhaps, most combustion engine cars today are equipped with the so-called lead-acid battery to start the vehicle (starter battery). This battery, utilizing fast electrode processes, was developed by Gaston Planté in 1859 and is still widely used: a remarkable example of an ingenious invention, which developed over centuries. It is even nowadays considered as one of the benchmarks that newly developed battery types are compared to [6]. Let us consider the working principles of this battery (Figure 6.5).

The battery consists of the high surface area metallic lead anode and lead oxide, PbO_2, cathode. Notably, PbO_2 is a good electronic conductor, enabling relatively fast kinetics of electrode redox processes and hence benefits the overall high current density. The spontaneous chemical reaction used in this device is simply the oxidation of metallic Pb by its oxide.

During the discharge, at the cathode, lead dioxide is spontaneously reduced to form lead sulfate, which remains in its solid state at the electrode surface, according to the following scheme:

$$PbO_2 + 4H^+ + SO_4{}^{2-} + 2e^- \rightarrow PbSO_4 + 2H_2O$$

At the anode side of this battery, the following electrode process takes place:

$$Pb + SO_4{}^{2-} - 2e^- \rightarrow PbSO_4$$

The overall redox reaction can be, therefore, written as follows:

$$Pb + PbO_2 + 2H_2SO_4 \rightarrow 2PbSO_4 + 2H_2O$$

The open-circuit voltage of the freshly charged new battery is *ca* 2.1 V. Interestingly, the chemical nature of the resulting deposits at the anode and the cathode is the same ($PbSO_4$) after discharge. This compound tends to undertake further irreversible chemical transformations with time.

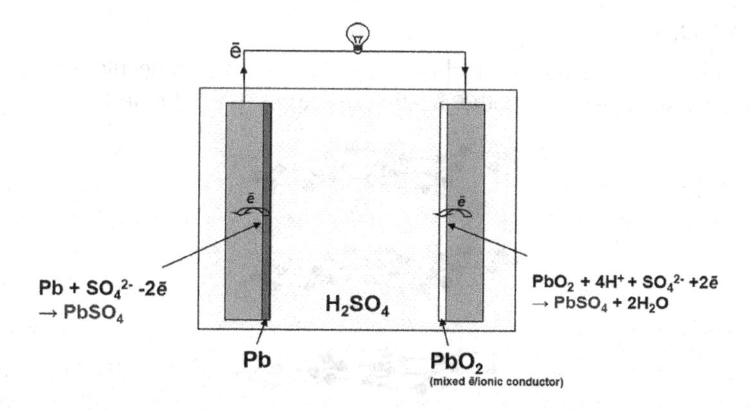

FIGURE 6.5 Schematics showing the principles of the operation of lead-acid batteries with metallic Pb anode and lead oxide cathode.

TABLE 6.2 Lead-Acid Batteries. Advantages vs. Disadvantages

Advantages	Disadvantages
• Inexpensive and simple to manufacture • Mature, reliable, and well-understood technology • Low self-discharge • Low maintenance requirements • Capable of high discharge rates	• Low volumetric and gravimetric energy densities • Cannot be stored in a discharged state • Allows only a relatively small number of complete discharge cycles • Toxic materials (Pb)

Therefore, it is not recommended to store lead-acid batteries in the discharged state to avoid losses in their available capacitance.

While lead-acid batteries have multiple advantages (see Table 6.2), including good availability of the materials, they have relatively low volumetric and gravimetric energy density. The use of concentrated sulfuric acid as the electrolyte further increases safety risks. These facts stimulated research in the field of batteries to involve light electroactive electrode materials to improve the abovementioned essential characteristics. Taking a look at the periodic table of elements, an obvious candidate would be metallic lithium or its compounds with other light elements. This is how the development of Li-ion batteries was rationalized and implemented.

6.4 RECHARGEABLE LI-ION BATTERIES

The Nobel Prize in Chemistry was awarded in 2019 to John Goodenough, Stanley Whittingham, and Akira Yoshino for their contributions to the development of rechargeable Li-ion batteries. Lithium-ion batteries make use of the phenomenon which is called *intercalation*.

DEFINITION:

Intercalation is the reversible inclusion or insertion of a molecule (or ion) into compounds with layered or other hosting structures (see Scheme 1).

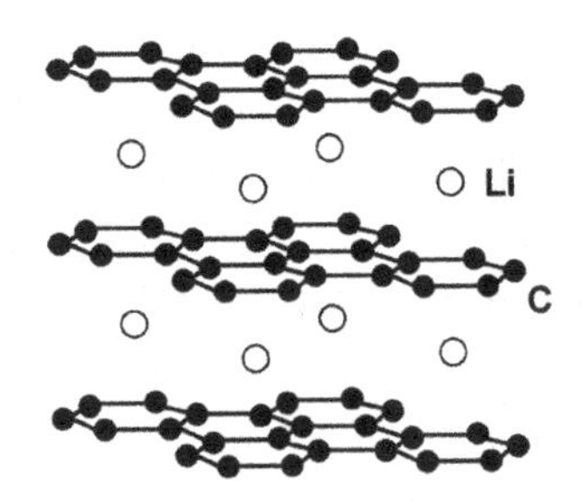

SCHEME 1 A schematic of the LiC_6 intercalation compound, where lithium is intercalated into the layered graphite structure.

The history of intercalation, which conceptually determines the design of materials for many state-of-the-art alkali metal-ion batteries, probably begins in 1840, when C. Schafhäutl observed graphite swelling in concentrated sulfuric acid [7]. However, only within few decades after the development of technical equipment for X-ray diffraction, the structures of the H_2SO_4/graphite and the alkali metal/graphite compounds were determined.

The term "intercalation" was probably used for the first time by McDonnell et al. in 1951 [8] (without any explanation for using it). In 1959, Rüdorff used the phrase "intercalation compounds" in the title of a review about all chemical derivatives of graphite [9]. In these compounds, atoms or ions can be inserted (alternatively, intercalated) under an expansion of the lattice perpendicular to the nearly unchanged graphite layers.

Probably the first battery using an intercalation compound was a battery (see Figure 6.6) with a titanium disulfide cathode and a lithium-aluminum anode (Stanley Whittingham, Nobel Prize winner 2019) [10]:

$$xLi + TiS_2 \rightarrow Li_xTiS_2$$

$$Li_xTiS_2(x<1)$$

A typical discharge curve of the first Li-TiS_2 battery is shown in Figure 6.7. Note that after the discharge, the stoichiometric amount of lithium does not reach 1.

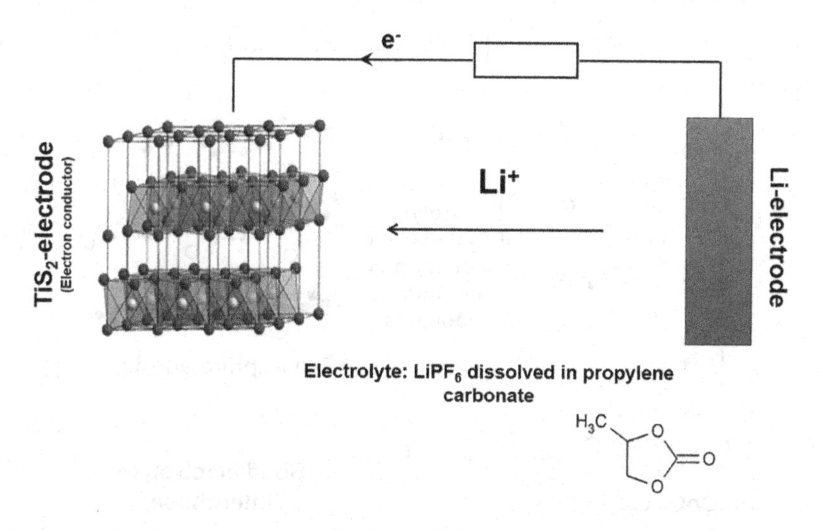

FIGURE 6.6 Schematic explaining the operation principles of one of the earliest Li-ion batteries.

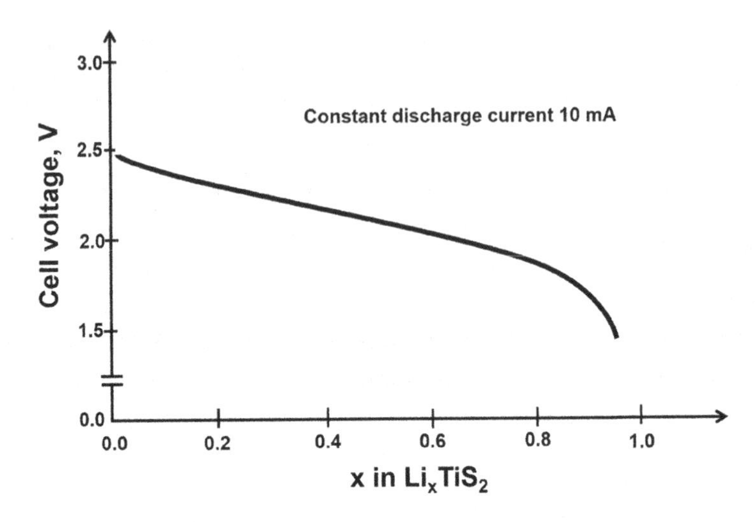

FIGURE 6.7 Discharge curve of one of the first Li-TiS_2 batteries. (Adapted from [10].)

Almost simultaneously with the work of Whittingham, electrochemical intercalation of lithium into graphite electrodes was investigated by Jürgen Besenhard [11]. These discoveries probably motivated another Nobel Prize winner, John Goodenough, to introduce graphite as an anode material and $LiCoO_2$ compound instead of TiS_2 as the cathode to suggest the battery, which we now know as the Li-ion battery (see Figure 6.8) [12,13].

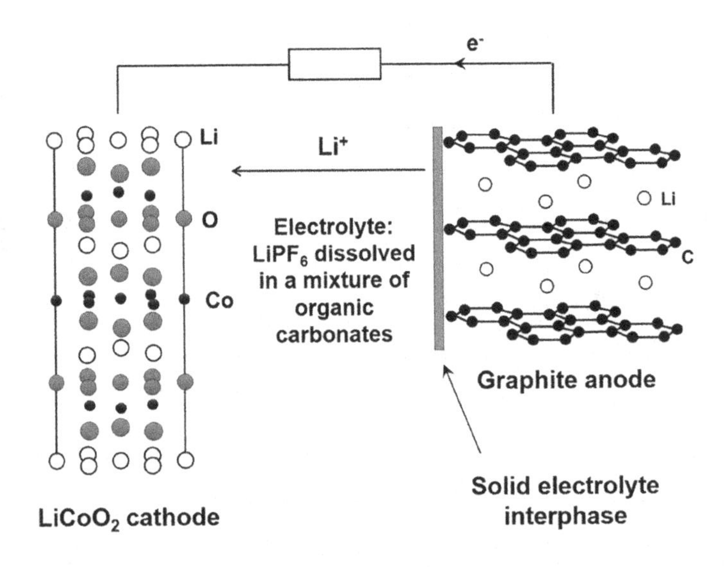

FIGURE 6.8 A schematic of a typical Li-ion battery.

The recent commercial success of Li-ion batteries stimulated further research to find a replacement of relatively expensive $LiCoO_2$ cathodes. Additionally, attempts were undertaken to find anode materials that would provide higher resulting battery capacities. Some alternatives to $LiCoO_2$ were suggested to be $LiFePO_4$ (olivine structure) and $LiMn_2O_4$ (spinel).

Lithium iron phosphate offers a longer cycle life. It is safe and reduces the cost and environmental concerns of $LiCoO_2$, particularly regarding cobalt entering the environment through improper disposal. The advantages of $LiMn_2O_4$ are high thermal stability, low cost, abundance, and environmental affinity. Some parameters characterizing the cathode materials of currently commercially successful Li-ion batteries are given in Table 6.3.

As for the anode material, the use of silicon electrodes [14,15] would increase the capacity gradually [16]. Compare the reaction taking place at the graphite anode:

$$6C + Li^+ + e^- \leftrightarrow LiC_6 \Rightarrow \sim 372\ \text{mAh/g}$$

with the reaction taking place at the silicon anode:

$$4Si + 15Li^+ + 15e^- \leftrightarrow Li_{15}Si_4 \Rightarrow \sim 3580\ \text{mAh/g}$$

However, alloying between Li and Si and, in general, the formation of Li-alloys [17] causes a significant change in the electrode dimensions (i.e., volume expansion up to ~400%) [18], leading to serious mechanical [19] and electrical consequences. Therefore, as a compromise, a small (several percent) addition of Si to the graphite anodes is currently used to moderately increase the battery capacity.

Typical Li-ion batteries use solutions of $LiPF_6$ salt in a mixture of non-aqueous ethylene carbonate and, e.g., propylene carbonate, dimethyl carbonate, diethyl carbonate, or ethyl-methyl carbonate solvents. One of

TABLE 6.3 Essential Characteristics of Cathode Materials for Commercial Li-ion Batteries

Cathode Material	Average Cathode Potential *vs* Li^+/Li, V	Specific Capacity, Ah/kg	Specific Energy, Wh/kg
$LiCoO_2$	~3.9	~140	~540
$LiNi_{0.33}Co_{0.33}Mn_{0.33}O_2$	~3.8	~170	~650
$LiMn_2O_4$	~4.1	~120	~490
$LiFePO_4$	~3.45	~170	~580

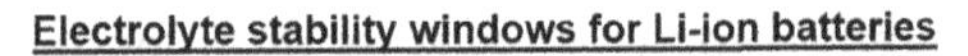

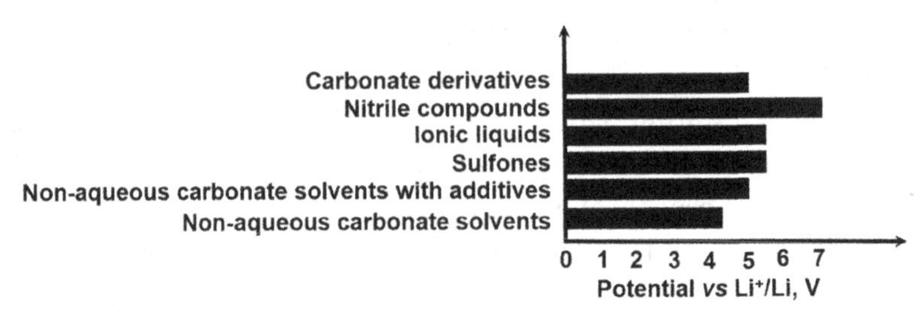

FIGURE 6.9 Electrochemical stability windows for some classes of solvents and electrolytes for Li-ion batteries.

the main problems with these solvents is that they are highly flammable. However, it is not easy to find a good alternative to them [20], which would enable a wide electrochemical stability window (> 4 V) required to maintain high battery voltage. Nevertheless, some candidates are recently considered promising. These classes of solvents are compared in Figure 6.9.

Another peculiarity related to the choice of solvents for Li-ion batteries is the formation of the so-called solid electrolyte interphase (SEI) at the anode (see Figure 6.8). An SEI appears at the surface of the anodes during the first few charging-discharging cycles as a result of partial decomposition of the electrolyte. The SEI passivates and thereby protects the anode surface and, at the same time, inhibits further electrolyte decomposition. The properties of these layers are under intensive investigation nowadays, as they determine battery cycling stability and performance.

The relatively low abundance of Li in the Earth's crust (only ~20 ppm) motivates replacing Li with Na (~26,000 ppm) or even K (~25,000 ppm) [21–24] in the batteries. Na^+ is, on the other hand, bigger than Li^+ by ~70%, determining the choice of the intercalation materials. Na-ion batteries have attracted significant attention in recent years [25]. However, due to the high energy density of Li-ion batteries, the potential markets for these batteries could be for the devices in which cycle life and cost are more essential factors than energy density. These use-cases include large-scale electricity storage for renewable forms of energy and electrical grids [26,27]. Interestingly, many electrode materials used in Li-ion batteries can also be utilized in sodium-ion batteries.

On the other hand, it should be noted that the intercalation of sodium into graphite happens mainly in the presence of specific organic solvents. Among them are the so-called *glymes*. Those are small-molecule linear and cyclic ethers like bis(2-methoxyethyl) ether (diglyme):

H_3CO O OCH_3

They have low viscosity and are able to coordinate with alkali metal ions, producing high concentrations of mobile charge carriers. One should also mention that Na-ion batteries utilizing ionic liquids demonstrate certain promise [28].

Besides, particularly interesting are sodium-ion batteries for stationary applications, which use aqueous electrolytes. These are briefly considered in the next section.

6.5 AQUEOUS NA-ION BATTERIES

Aqueous sodium-ion batteries are a class of energy storage devices that use relatively inexpensive and safe aqueous electrolytes with dissolved salts like Na_2SO_4 or $NaClO_4$ and electrodes typically made of affordable materials. Recent interest in these batteries (as well as in aqueous K-ion batteries [29]) is due to their possible applications in renewable energy provision as large-scale storage systems [30,31]. Typical electrode materials for Na-ion batteries are compared in Figure 6.10 without

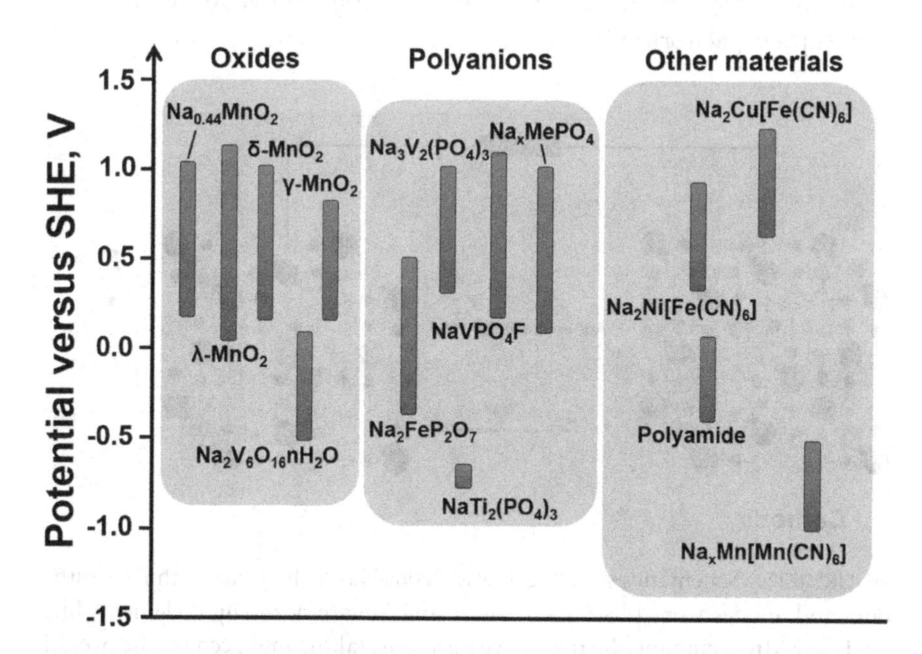

FIGURE 6.10 Working electrode potentials (versus standard hydrogen reference electrode) for different classes of electrode materials for aqueous Na-ion batteries. (The data are from [35,36].)

considering organic electrode materials [32]. From this Figure, one can see that there is an attractive class of materials with the general chemical formula $Na_xMe[Me^{host}(CN)_6]$, the so-called Prussian blue analogs (PBAs), where Me and Me^{host} designate *d*-elements like Ni, Cu, or Mn. Depending on these elements' chemical nature, the resulting PBAs can work as either anode or cathode materials [33,34]. Nevertheless, it is also clear that there is a very limited choice of materials for the anodes compared to the cathodes.

Figure 6.11 schematically shows the working principles of aqueous Na-ion batteries. It illustrates a situation when the anode and cathode are made of the PBA-analog materials, which are relatively good electron conductors. PBAs are recently considered state-of-the-art materials for this class of aqueous batteries. These materials have a crystal structure to allow the intercalation of almost all alkali metal cations.

The overall capacity of the PBA-electrodes does not normally exceed 60 Ah/kg, which is slightly larger than the capacity of lead-acid batteries. This is particularly due to the limited stability of aqueous electrolytes compared to the organic ones [37]. However, these materials often enable very fast charge, similar or comparable to the situation with supercapacitors [38].

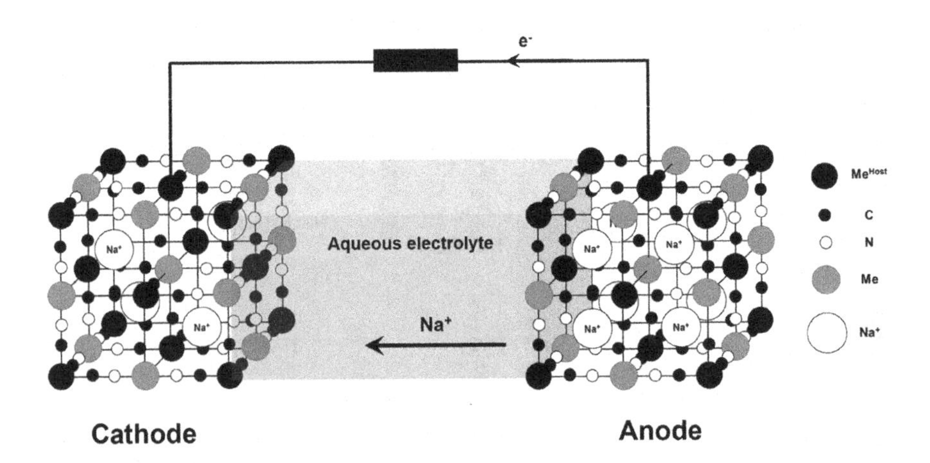

FIGURE 6.11 Schematics showing how aqueous Na-ion batteries with PBA cathodes and anodes work (discharge). Me^{host} and Me are normally d-elements like Ni, Fe, or Mn occupying the respective positions, taking into account the overall formula $Na_2Me[Me^{host}(CN)_6]$.

6.6 METAL-AIR BATTERIES

Metal-air batteries practically combine the ideas behind the working principles of classical batteries and fuel cells. At the anode, they have a metal, such as zinc [39,40], and at the cathode, there are composite electrodes with a bifunctional catalyst promoting both the oxygen reduction and oxygen evolution reactions. Moreover, the cathode should prevent the leakage of the electrolyte and at the same time enable the transfer of oxygen from the environment to the catalyst surface. A schematic of the working principles of Zn-air batteries is shown in Figure 6.12.

As one can see, the anode resembles a part of a classical battery system, where the reaction product remains in the system. On the other hand, the cathode looks like the cathode of fuel cells, where the oxygen reduction reaction occurs during the discharge.

The main advantage of this battery is that all the functional materials are affordable and nontoxic. Additionally, the fact that the oxidant can be taken from the atmosphere increases the gravimetric energy density of the device. On the other hand, there are some challenges [41], for instance,

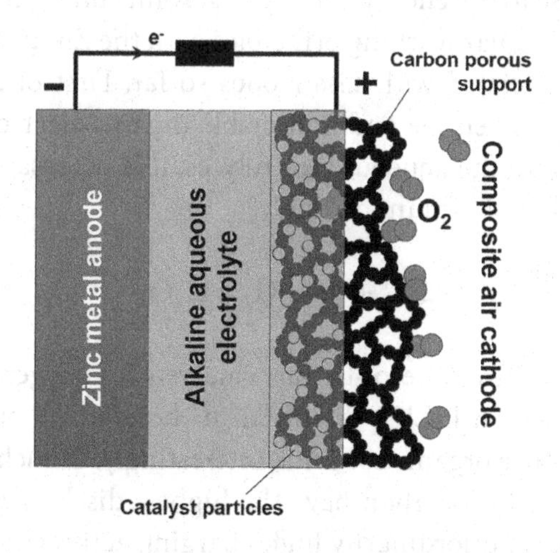

FIGURE 6.12 Aqueous zinc-air battery: an explanation of the principles. During the discharge, the metal anode is oxidized, and the resulting Zn^{2+} forms the complex $[Zn(OH)_4]^{2-}$soluble in alkaline aqueous solutions. At the cathode, oxygen molecules from air are reduced at the catalyst particles supported by porous carbon-based materials. During the charge, the zinc complex is chemically reduced back to metallic Zn. At the same time, the catalyst at the opposite side should enable efficient catalysis of the oxygen evolution reaction.

related to the formation of the so-called dendrites at the Zn-electrode side during charging, which decreases the battery's lifetime. Re-deposition of zinc metal in a needle-like structure, called a dendrite, results in the short-circuiting of the system. Another challenge is the catalytic material of the cathode. As mentioned, it should be bifunctional [42]. In other words, it should catalyze both the oxygen reduction reaction during the discharging and the oxygen evolution reaction during charging, which is hard to design [43]. As one can see from the previous discussions on electrocatalysts, the best catalysts for the oxygen reduction reaction undergo chemical transformation upon anodic polarization to become often only moderate catalysts for the oxygen evolution reaction and *vise versa*.

While Zn-air batteries are promising as probably the most affordable (from the point of view of used materials) energy storage devices, other metal-air batteries also attract considerable attention. For instance, if one uses metallic lithium as the anode material, one can theoretically achieve a much higher specific energy density [44]. For the charged state, if oxygen is excluded, the specific energy density of such a battery is *ca* 11680 Wh kg^{-1}. This value is close to the energy density of gasoline under comparable conditions. Despite similar working principles with the Zn-air batteries, other problems are associated with Li-air ones so far. First of all, in the case of lithium, more expensive and flammable organic electrolytes are often used. In contrast to the aqueous electrolytes, it is necessary to catalyze at least the formation of lithium peroxide:

$$2Li^+ + 2e^- + O_2 = Li_2O_2$$

Platinum and gold are state-of-the-art catalysts for oxygen reduction and evolution reactions in $LiClO_4$ dissolved in the mixture of the propylene carbonate and other organic solvents. Interestingly, in such batteries, gold particles supported on carbon have the highest discharge activity, while the Pt/C exhibits extraordinarily high charging activity. Pt/Au/C materials show bifunctional catalytic activity for oxygen reduction and evolution reaction kinetics in lithium-oxygen cells.

There are also important requirements for the electrolyte in the non-aqueous lithium-air system, which are as follows:

1. Stable in contact with lithium metal
2. High oxidation/reduction potentials to avoid electrolyte decomposition

3. Low vapor pressure and high boiling point
4. High lithium salt solubility and good chemical stability

Historically, the first electrolyte was a gel-type polymer electrolyte of $LiPF_4$ in polyacrylonitrile with ethylene carbonate and propylene carbonate, which is considered quite dangerous. Nowadays, the big hope is to use ionic liquids to overcome this issue. Lithium-air battery research and technology are still in their initial stage. The technical basis for practical high power density and extended deep cycling has yet to be demonstrated.

There is another class of metal-air batteries, namely mechanically rechargeable ones. A good example is Al-air devices, where metallic aluminum is used as an anode [45]. The estimated gravimetric energy density of such a battery is *ca* 8100 Wh kg^{-1} (see Table 6.4). This ranks it as number two after the Li-air systems and makes it attractive for, e.g., automotive applications. The idea is that after the discharge that produces aluminum hydroxide, $Al(OH)_3$, or oxide, Al_2O_3, one can physically replace the batteries quickly with the new fresh ones at numerous refueling stations. The discharged battery is then transferred to the mechanical recharge, where the, e.g., concentrated thermal solar energy is used to reduce the aluminum oxide/hydroxide back to the metallic aluminum. Several pilot projects demonstrated that this technology could be viable for automotive applications if a suitable infrastructure were built.

6.7 ALL-SOLID-STATE BATTERIES

Commercial rechargeable Li-ion batteries, as it is clear now, approach their theoretical capacity limits. Therefore, attempts are undertaken to use the lightest metal, lithium, as an anode in contact with solid electrolytes to increase the device's gravimetric capacity (all-solid-state batteries) [46,47].

TABLE 6.4 Estimated Energy Densities of Some Metal-Air Batteries

Battery Type	Estimated Energy Density, kWh kg^{-1}
Li-air	~11.7
Al-air	~8.1
Mg-air	~6.8
Ca-air	~4.2
Na-air	~2.3
K-air	~1.7
Zn-air	~1.3

This would also potentially solve many problems related to safety (no flammable electrolyte) and dendrite formation and help to enable comparably simple fabrication of the battery.

One idea is to use solid (ceramic) lithium ionic conductors [48] to replace the organic electrolytes, so that metallic Li anodes and probably classical cathode materials like lithium metal oxides or similar can be used [49–53]. However, while the overall idea is very attractive, there are also considerable challenges on the way to a successful commercial device, e.g., for automotive applications. First of all, the electrolyte should be stable in contact with metallic lithium anodes [54,55], a situation which is so far difficult to achieve. There might be difficulties related to the formation of additional mass and charge transfer barriers between the electrodes and electrolytes due to the so-called space charge layers [56], where some parts of the system demonstrate excessive charge as a result of the electrochemical process or due to simple polarization [57]. Furthermore, a good solid electrode/solid electrolyte contact should be achieved [58] with minimal effort. Further research should demonstrate if this idea is a new step in developing batteries for portable and automotive applications (Figure 6.13).

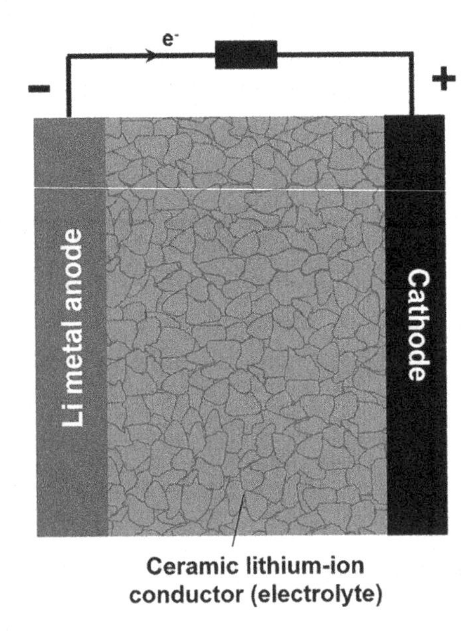

FIGURE 6.13 A general concept of the Li-ion all-solid-state battery, with metallic lithium as the anode material and ceramic Li-ion conducting solid electrolytes.

6.8 SUMMARY AND CONCLUSIONS

Efficient functional materials for various battery applications (for both primary and secondary units) are of enormous significance today for portable and automotive applications and are critical for future sustainable energy provision.

State-of-the-art primary batteries should be free from toxic and scarce materials due to difficulties in recycling massive amounts of primary units worldwide annually (safety and economic aspects are decisive). The primary batteries will most likely be useful mainly for portable applications in the future.

For larger-scale renewable energy applications, secondary batteries are more useful. Li-ion batteries are state-of-the-art rechargeable systems for portable and automotive applications. However, other strategies to improve the performance of this type of battery are necessary as they approach the theoretical capacity maximum for the current materials used in these devices. Overcoming problems that are mechanical in nature should also be considered, i.e., changes in physical dimensions of the electrodes during intercalation and de-intercalation.

Na-ion batteries have also attracted significant attention in recent years. However, due to the high energy density of Li-ion batteries, the potential markets for these batteries are large-scale stationary applications. Moreover, aqueous Na-ion batteries enable better safety and lower costs of the electrolyte components.

Metal-air batteries are also considered promising. Viable approaches can include abundant materials, for example, Zn in Zn-air batteries. For automotive applications, Li-air and mechanically rechargeable Al-air batteries seem to be promising. However, lithium-air battery research and technology are still in their initial stage. The technical basis for practical high power density and extended deep cycling has yet to be demonstrated.

6.9 QUESTIONS

1. Define a primary cell/battery and a secondary cell/battery. Analyze what the role and capacity of these batteries vs. other energy provision devices is.

2. Explain the working principles of AMDBs. What are state-of-the-art materials for these batteries? Analyze the advantages vs. disadvantages of such primary batteries.

3. Explain the working principles of a lead-acid battery.
4. What are the working principles and materials used in zinc-air batteries? Analyze the advantages and disadvantages of zinc-air batteries (and issues with materials for them).
5. Why is it promising to use Zn-air and Al-air batteries for automotive applications?
6. Define the intercalation process. What are typical intercalation compounds? Name state-of-the-art materials for Li-ion batteries.
7. What was the first battery using an intercalation compound?
8. What are the working principles of Li-ion batteries? What are the typical cathode and anode materials for Li-ion battery applications?
9. Analyze further possible directions for the development of Li-ion batteries. Why is silicon a promising material to replace graphite in Li-ion batteries?
10. What are the working principles of aqueous and nonaqueous Na-ion batteries? Name the state-of-the-art electrode materials for these devices. Analyze the promises and challenges in the development of Li-air batteries.

REFERENCES

1. Armand, M.; Tarascon, J.-M. 2008. Building better batteries. *Nature* 451:652–657.
2. Durmus, Y.E.; Zhang, H.; Baakes, F.; Desmaizieres, G.; Hayun, H.; Yang, L.; Kolek, M.; Küpers, V.; Janek, J.; Mandler, D.; Passerini, S.; Ein-Eli, Y. 2020. Side by side battery technologies with lithium-ion based batteries. *Advanced Energy Materials* 24:2000089.
3. Sánchez-Díez, E.; Ventosa, E.; Guarnieri, M.; Trovò, A.; Flox, C.; Marcilla, R.; Soavi, F.; Mazur, P.; Aranzabe, E.; Ferret, R. 2021. Redox flow batteries: Status and perspective towards sustainable stationary energy storage. *Journal of Power Sources* 481:228804.
4. Zhang, J.; Jiang, G.; Xu, P.; Kashkooli, A.G.; Mousavi, M.; Yu, A.; Chen, Z. 2018. An all-aqueous redox flow battery with unprecedented energy density. *Energy & Environmental Science* 11:2010–2015.
5. Marsal, P.A.; River, R.; Kordesch, K.; Lakewood; Urry, L.F. Dry cell. *United States Patent* 2960558.
6. Lopes, P.P.; Stamenkovic, V.R. 2020. Past, present, and future of lead–acid batteries. *Science* 369:923–924.

7. Schafhäutl, C. 1840. Über die Verbindungen des Kohlenstoffes mit Silicium, Eisen und anderen Metallen, welche die verschiedenen Arten von Gusseisen, Stahl und Schmiedeeisen bilden. *Journal für Praktische Chemie* 21:159–174.
8. McDonnell, F.R.M.; Pink, R.C.; Ubbelohde, A.R. 1951. Some physical properties associated with "aromatic" electrons. Part III. The pseudo-metallic properties of potassium–graphite and graphite–bromine. *Journal of Chemical Society* 1951:191–197.
9. Rüdorff, W. 1959. Graphite intercalation compounds. *Advances in Inorganic Chemistry and Radiochemistry* 1:223–266.
10. Whittingham, S. 1976. Electrical energy storage and intercalation chemistry. *Science* 192:1126–1127.
11. Besenhard, J.O. 1976. The electrochemical preparation and properties of ionic alkali metal-and NR4-graphite intercalation compounds in organic electrolytes. *Carbon* 14:111–115.
12. Goodenough, J.B. 2018. How we made the Li-ion rechargeable battery. *Nature Electronics* 1:204.
13. Xie, J.; Lu, Y.C. 2020. A retrospective on lithium-ion batteries. *Nature Communications* 11:2499.
14. Obrovac, M.N.; Chevrier, V.L. 2014. Alloy negative electrodes for Li-ion batteries. *Chemical Reviews* 114:11444–11502.
15. Liu, N.; Lu, Z.; Zhao, J.; McDowell, M.T.; Lee, H.W.; Zhao, W.; Cui, Y. 2014. A pomegranate-inspired nanoscale design for large-volume-change lithium battery anodes. *Nature Nanotechnology* 9:187–192.
16. Wu, H.; Cui, Y. 2012. Designing nanostructured Si anodes for high energy lithium ion batteries. *Nano Today* 7:414–429.
17. Corsi, J.S.; Welborn, S.S.; Stach, E.A.; Detsi. E. 2021. Insights into the degradation mechanism of nanoporous alloy-type Li-ion battery anodes. *ACS Energy Letters* 6:1749–1756.
18. Obrovac, M.N.; Christensen, L. 2004. Structural changes in silicon anodes during lithium insertion/extraction. *Electrochemical Solid State Letters* 7:A93–A96.
19. Deshpande, R.; Cheng, Y-T.; Verbrugge, M.W. 2010. Modeling diffusion-induced stress in nanowire electrode structures. *Journal of Power Sources* 195:5081–5088.
20. Yang, Q.; Zhang, Z.; Sun, X.G.; Hu, Y.S.; Xing, H.; Dai, S. 2018. Ionic liquids and derived materials for lithium and sodium batteries. *Chemical Society Reviews* 47:2020–2064.
21. Anoopkumar V; John, B.; Mercy T.D. 2020. Potassium-ion batteries: key to future large-scale energy storage? *ACS Applied Energy Materials* 3:9478–9492.
22. Min, X.; Xiao, J.; Fang, M.; Wang, W.; Zhao, Y.; Liu, Y.; Abdelkader, A.M.; Xi, K.; Kumar, R.V.; Huang, Z. 2021. Potassium-ion batteries: Outlook on present and future technologies. *Energy & Environmental Science* 14:2186–2243.
23. Hosaka, T.; Kubota, K.; Hameed, A.S.; Komaba, S. 2020. Research development on K-ion batteries. *Chemical Reviews* 120:6358–6466.

24. Dhir, S.; Wheeler, S.; Capone, I.; Pasta, M. 2020. Outlook on K-ion batteries. *Chem* 6:2442–2460.
25. Hwang, J.Y.; Myung, S.T.; Sun, Y.K. 2017. Sodium-ion batteries: present and future. *Chemical Society Reviews* 46:3529–3614.
26. Liu, T.; Zhang, Y.; Jiang, Z.; Zeng, X.; Ji, J.; Li, Z.; Gao, X.; Sun, M.; Lin, Z.; Ling, M.; Zheng, J.; Liang, C. 2019. Exploring competitive features of stationary sodium ion batteries for electrochemical energy storage. *Energy & Environmental Science* 12:1512–1533.
27. Yang, C.; Xin, S.; Mai, L.; You, Y. 2021. Materials design for high-safety sodium-ion battery. *Advanced Energy Materials* 11:2000974.
28. Matsumoto, K.; Hwang, J.; Kaushik, S.; Chen, C.Y.; Hagiwara, R. 2019. Advances in sodium secondary batteries utilizing ionic liquid electrolytes. *Energy & Environmental Science* 12:3247.
29. Pasta, M.; Wessells, C.D.; Huggins, R.A.; Cui, Y. 2012. A high-rate and long cycle life aqueous electrolyte battery for grid-scale energy storage. *Nature Communications* 3:1–7.
30. Pasta, M.; Wessells, C.D.; Liu, N.; Nelson, J.; McDowell, M.T.; Huggins, R.A.; Toney, M.F.; Cui, Y. 2014. Full open-framework batteries for stationary energy storage. *Nature Communications* 5:1–9.
31. Wang, R.Y.; Shyam, B.; Stone, K.H. Nelson Weker, J.; Pasta, M.; Lee, H.W.; Toney, M.F.; Cui, Y. 2015. Reversible multivalent (monovalent, divalent, trivalent) ion insertion in open framework materials. *Advanced Energy Materials* 5:1401869.
32. Han, C.; Zhu, J.; Zhi, C.; Li, H. 2020. The rise of aqueous rechargeable batteries with organic electrode materials. *Journal of Materials Chemistry A* 8:15479–15512.
33. Qian, J.; Zhou, M.; Cao, Y.; Ai, X.; Yang, H. 2012. Nanosized $Na_4Fe(CN)_6$/C Composite as a low-cost and high-rate cathode material for sodium-ion batteries. *Advanced Energy Materials* 2:410–414.
34. Hurlbutt, K.; Wheeler, S.; Capone, I.; Pasta, M. 2018. Prussian blue analogs as battery materials. *Joule* 2:1950–1960.
35. Kim, H.; Hong, J.; Park, K.Y.; Kim, H.; Kim, S.W.; Kang, K. 2014. Aqueous rechargeable Li and Na ion batteries. *Chemical Reviews*, 114:11788– 11827.
36. Yun, J.; Schiegg, F.; Liang, Y.; Scieszka, D.; Garlyyev, B.; Kwiatkowski, A.; Wagner, T.; Bandarenka, A.S. 2018. Electrochemically formed $Na_xMn[Mn(CN)_6]$ thin film anodes demonstrate sodium intercalation and de-intercalation at extremely negative electrode potentials in aqueous media. *ACS Applied Energy Materials* 1:123–128.
37. Liu, Z.; Huang, Y.; Huang, Y.; Yang, Q.; Li, X.;Huang, Z.; Zhi, C. 2020. Voltage issue of aqueous rechargeable metal-ion batteries. *Chemical Society Reviews* 49:180–232.
38. Yun, J.; Pfisterer, J.; Bandarenka, A.S. 2016. How simple are the models of Na-intercalation in aqueous media? *Energy & Environmental Science* 9:955–961.
39. Li, Y.; Dai, H. 2014. Recent advances in zinc–air batteries. *Chemical Society Reviews* 43:5257–5275.

40. Gu, P.; Zheng, M.; Zhao, Q.; Xiao, X.; Xue, H.; Pang, H. 2017. Rechargeable zinc-air batteries: A promising way to green energy. *Journal of Materials Chemistry A* 5:7651–7666.
41. Zhang, J.; Zhou, Q.; Tang, Y.; Zhang, L.; Li, Y. 2019. Zinc-air batteries: Are they ready for prime time? *Chemical Science* 10:8924–8929.
42. Davari, E.; Ivey, D.G. 2018. Bifunctional electrocatalysts for Zn–air batteries. *Sustainable Energy and Fuels* 2:39–67.
43. Aijaz, A.; Masa, J.; Rösler, C.; Xia, W.; Weide, P.; Botz, A.J.R.; Fischer, R.A.; Schuhmann, W.; Muhler, M. 2016. Co@Co_3O_4 encapsulated in carbon nanotube-grafted nitrogen-doped carbon polyhedra as an advanced bifunctional oxygen electrode. *Angewandte Chemie International Edition* 55:4087–4091.
44. Bruce, P.G.; Freunberger, S.A.; Hardwick, L.J.; Tarascon, J.-M. 2012. Li-O_2 and Li-S batteries with high energy storage. *Nature Materials* 11:19–29.
45. Liu, Y.; Sun, Q.; Li, W.; Adair, R.K.; Li, J.; Sun, X. 2017. A comprehensive review on recent progress in aluminum–air batteries. *Green Energy & Environment* 2:246–277.
46. Krauskopf, T.; Richter, F.H.; Zeier, W.G.; Janek, J. 2020. Physicochemical concepts of the lithium metal anode in solid-state batteries. *Chemical Reviews* 120:7745–7794.
47. Randau, S.; Weber, D.A.; Kotz, O.; Koerver, R.; Braun, V.; Weber, A.; Ivers-Tiffee, E.; Adermann, T.; Kulisch, J.; Zeier, W.G.; Richter, F.H.; Janek, J. 2020. Benchmarking the performance of all-solid-state lithium batteries. *Nature Energy* 5:259–270.
48. Janek, J.; Zeier, W.G. 2016. A solid future for battery development. *Nature Energy* 1:16141.
49. Kim, K.J.; Balaish, M.; Wadaguchi, K.; Rupp, J.L.M. 2021. Solid-state Li–metal batteries: Challenges and horizons of oxide and sulfide solid electrolytes and their interfaces. *Advanced Energy Materials* 11:2002689.
50. Zhu, Y.; Gonzalez-Rosillo, J.C.; Balaish, M.; Hood, Z.D.; Kim, K.J.; Rupp, J.L.M. 2021. Lithium-film ceramics for solid-state lithionic devices. *Nature Reviews Materials* 6:313–331.
51. Balaish, M.; Gonzalez-Rosillo, J.C.; Kim, K.J.; Zhu, Y.; Hood, Z.D.; Rupp, J.L.M. 2021. Lithium-film ceramics for solid-state lithionic devices. *Nature Energy* 6:227–239.
52. Kim. K.J.; Hinricher, J.J.; Rupp, J.L.M. 2020. High energy and long cycles. *Nature Energy* 5:278–279.
53. Lee, Y.G.; Fujiki, S.; Jung, C.; Suzuki, N.; Yashiro, N.; Omoda, R.; Ko, D.S.; Shiratsuchi, T.; Sugimoto, T.; Ryu, S.; Ku, J.H.; Watanabe, T.; Park, Y.; Aihara, Y.; Im, D.; Han, T. 2020. High-energy long-cycling all-solid-state lithium metal batteries enabled by silver–carbon composite anodes. *Nature Energy* 5:299–308.
54. Chen, R.S.; Li, Q.H.; Yu, X.Q.; Chen, L.Q.; Li, H. 2020. Approaching practically accessible solid-state batteries: stability issues related to solid electrolytes and interfaces. *Chemical Reviews* 120:6820–6877.
55. Zhu, Y.; Connell, J.G.; Tepavcevic, S.; Zapol, P.; Garcia-Mendez, R.; Taylor, N.J.; Sakamoto, J.; Ingram, B.J.; Curtiss, L.A.; Freeland, J.W.; Fong, D.D.;

Markovic, N.M. 2019. Dopant-dependent stability of garnet solid electrolyte interfaces with lithium metal. *Advanced Energy Materials* 9:1803440.

56. Katzenmeier, L.; Carstensen, L.; Schaper, S.J.; Müller-Buschbaum, P.; Bandarenka, A.S. 2021. Characterization and quantification of depletion and accumulation layers in solid-state Li^+-conducting electrolytes using in situ spectroscopic ellipsometry. *Advanced Materials* 33:2100585.
57. Katzenmeier, L.; Helmer, S.; Braxmeier, S.; Knobbe, E.; Bandarenka, A.S. 2021. Properties of the space charge layers formed in Li-ion conducting glass ceramics. *ACS Applied Materials and Interfaces* 13:5853–5860.
58. Hänsel, C.; Kundu, D. The stack pressure dilemma in sulfide electrolyte based Li metal solid-state batteries: A case study with Li_6PS_5Cl Solid Electrolyte. *Advanced Materials Interfaces* 2021:2100206.

CHAPTER 7

An "Artificial Leaf": A Dream or a Viable Energy Provision Concept?

7.1 MOTIVATION

It is a known fact that our planet receives approximately 5000 times more energy from the sun than humankind currently uses in a year. Therefore, the energy received in about 2 hours of sunlight would be enough to cover our annual energy consumption. Would it be possible to "harvest" it by mimicking natural processes, for example, photosynthesis [1,2]? The answer today can be rather optimistic. However, mimicking the biological processes is not trivial due to the complexity of the bio-systems and the processes taking place there. For instance, even the initial steps, which occur during photosynthesis in the bio-systems, briefly and roughly described below, illustrate the challenges of creating exact analogs of such objects [3].

Photosynthesis starts in the so-called photosystems I and II [4], which absorb the 700 and 680 nm photons, respectively. They use these photons to oxidize H_2O, with a quantum yield close to unity. The whole initial process may be conceptually divided into three key steps.

DOI: 10.1201/9781003025498-7

1. Initial light absorption occurs, followed by local charge separation
2. Proton-coupled electron transfer between redox cofactors (i.e., redox-active components of catalytic enzymes) along the photosynthetic chain allows further spatial charge separation and prevents charge recombination
3. Multielectron generation of H-containing intermediates and oxygen happens. It is catalyzed by enzymatic sites, like the bi-nuclear Ni-Fe and Fe-Fe clusters in hydrogenases or oxygen-evolving Ca-Mn_4 centers of the photosystem II [4,5]

However, one can split the complex natural systems and processes into much simpler fragments and steps, which might be easier to understand and mimic using various concepts. For this reason, only steps 1 and 2 mentioned above have been the core of the research in artificial photosynthesis during the last few decades worldwide.

While natural photosynthesis aims to form natural fuels, i.e., sugars, one can also focus on producing just gaseous hydrogen [6,7] or other small molecules [8,9], not sugars. Some microorganisms, such as cyanobacteria or microalgae, are able to photosynthesize H_2 under very specific conditions. Why not mimic this particular process? Indeed, producing a fuel like H_2 using just sunlight seems attractive and viable, and this idea is under consideration for decades [10–12].

DEFINITION:

Artificial photosynthesis aims to mimic natural photosynthesis by engineering schemes, which can produce practical (for the current world economy) fuels, like hydrogen, by splitting water using sunlight.

What is essential is that one needs to use relatively straightforward strategies: artificial photosynthesis should increase the energy conversion efficiency and the long-term stability of the fuel generating systems. This can be done by replacing biological agents in the natural schemes, which are considered unstable from the industrial point of view, with, e.g., durable solid-state inorganic or organic materials.

In principle, approaches for artificial photosynthesis demonstrate an unprecedented diversity in the ways humankind addresses these challenges [13]. For example, one might significantly simplify the systems using semiconductor materials [14,15]. Alternatively, one can try to immobilize naturally existing *stable* objects at the surface of conducting solids and directly integrate them into the currently used energy provision schemes [16–18]. It could be attractive to use some molecular complexes as water splitting systems in mainly aqueous solutions under sunlight illumination [19–23]. Additionally, it is feasible to perform the so-called photoelectrochemical water splitting [24–26], where no or relatively small additional external voltage is necessary. In the following, some of these approaches are considered in more detail.

7.2 A PHOTOCATALYTIC APPROACH

Probably the most straightforward concept to perform artificial photosynthesis can be formulated as follows. One just uses a *photocatalyst* dispersed in an aqueous electrolyte [27–30]. This material helps to split water to H_2 and O_2 under illumination without any other additional material, neither electronically conducting nor isolating.

DEFINITION:

A photocatalyst is a material that gets catalytically activated (alters the rate of a chemical reaction) only when exposed to the (sun)light.

In the current practice, photocatalysts are typically semiconductor materials [31]. To understand why they are attractive, let us consider the processes taking place in such semiconductor systems. For that, we will begin with the so-called energy diagrams of semiconductors. For simplicity, we will not consider the effects that occur at the interface between different types of conductors and materials with other semiconducting properties (i.e., p-doped and n-doped semiconductors, discussed in later sections).

The electronic conductivity of solids is now well-understood using the so-called band theory of solids. Using that theory, one can explain why some solids are insulators, whereas others are good electron conductors. Below is a summary of the output of this theory.

There are, in principle, two approaches in that theoretical framework to describe the electron energies in a periodic lattice of atoms. The first approach involves free electrons affected by a periodic potential (caused by lattice ions). The second approach considers isolated atoms, which are brought together to form solids. Both approximations give similar results: discrete energy levels, which are allowed for electrons, are grouped, forming *energy bands*.

As it is known from quantum mechanics, two atoms can create a joint system. In the simplest case, its stability will be determined by the energy levels allowed for the electrons belonging to the system and the occupation of these levels with the electrons. Suppose one considers a two-atom system (for example, the hydrogen molecule, H_2) with only one electron contributing from each side. In that case, one can expect the formation of the so-called bonding and antibonding energy levels or orbitals (Figure 7.1). The antibonding levels are located much higher in the energy scale. The two electrons in a hydrogen molecule occupy the bonding molecular orbital in the ground state, which corresponds to the highest probability of finding the electrons between the positively charged nuclei. However, one can excite the electrons by providing, for instance, photons with the right energy to bring them to the antibonding molecular orbital. This would correspond to zero probability to find an electron at the midpoint between the atoms; instead, there is an increased probability of finding the electron outside the region between two nuclei.

Similarly, if one brings, for example, ten atoms of the same nature close together, one can assume that the orbitals might group as schematically shown in Figure 7.2. While it is still possible to distinguish different energy levels within these groups, the difference between them can be so small that electrons can jump between them with relative ease just due to a small energy influx. In the ground state, the levels, which are lower in energy, are occupied with the electrons, while the upper ones remain empty. In the lower levels, therefore, the electrons are "bound". In the empty upper levels, they acquire more degrees of freedom to occupy

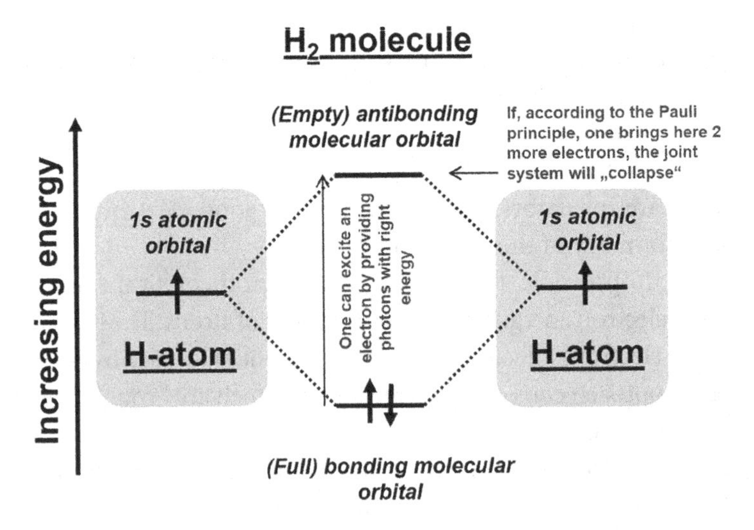

FIGURE 7.1 Formation of bonding and antibonding molecular orbitals in a hydrogen molecule.

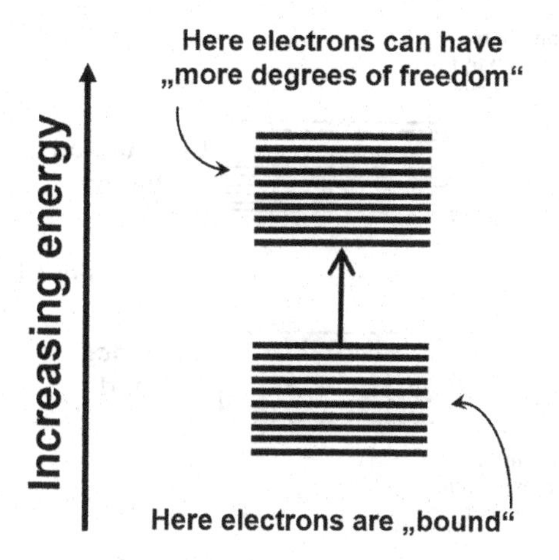

FIGURE 7.2 Schematic energy diagram for the case of ten atoms of the same kind in close proximity.

different neighboring orbitals. This is equivalent to the higher mobility of these electrons compared to the bound electrons.

The formation of multiatomic systems results in the appearance of the numerous allowed energy levels for the charged species that it is practically difficult to distinguish them. In this case, one can consider them as continuous energy *bands*. The highest band occupied with the electrons is called the *valence band*. Consequently, the lowest unoccupied band is called the *conduction band*, and the electrons with these energies will have increased mobility. Those bands overlap for many metals explaining their high electronic conductivity even at low temperatures. However, for semiconductors (and dielectrics), there is a significant energy gap between the valence and the conduction bands. Therefore, it is less likely to find electrons in the conduction bands; this explains the low electronic conductivity of such materials (Figure 7.3).

DEFINITION:

The band gap (energy), often designated as E_g, is the smallest separation of the valence and conduction bands.

In the model shown in Figure 7.3, it is also essential to distinguish the so-called Fermi level, E_F, or the electrochemical potential of electrons. According to Fermi-Dirac's statistics, E_F expresses the density and the average energy of the quantum states of electrons, which are in reality in

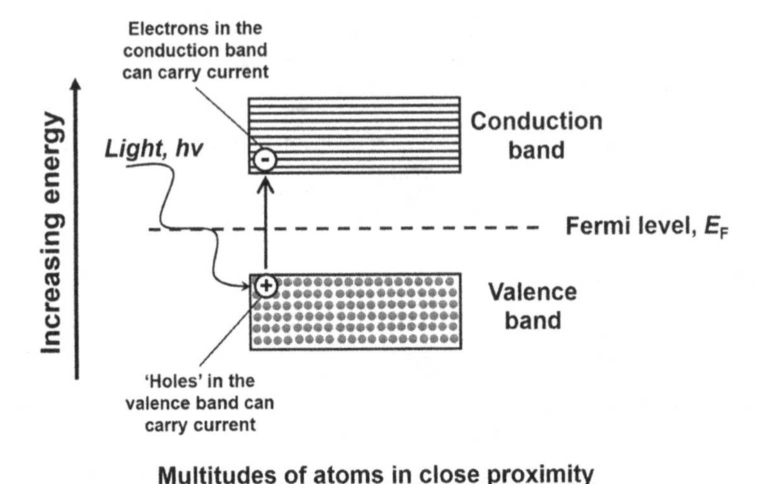

FIGURE 7.3 Schematic energy diagram for intrinsic semiconductors.

the conduction and valence bands. The closer the Fermi level is to the conduction band, the higher the probability of finding electrons there.

It is possible to excite electrons from the valence band to the conduction band if, e.g., a photon with energy greater than E_g is absorbed. In this case, an *electron-hole pair* is formed (see Figure 7.3). The *hole* can be considered as a sort of defect in, e.g., a covalent bond. Both electrons and holes have increased mobility and can contribute to the electronic conductivity of solids. One should note that the electron-hole pairs are also constantly generated due to thermal energy and any external energy sources.

Another step to understanding the models used in this field of energy materials science is to introduce the relationship between standard physical and electrochemical energy scales (Figure 7.4). As one can see, those scales have opposite directions. Additionally, there is a difference between them of 4.44 ± 0.02 eV at 298.15 K (if SHE scale is used) [32].

Now we are ready to construct models used to explain how one can use semiconductors to organize artificial photosynthesis (Figure 7.5). Let us start with a schematic drawing of a photocatalyst particle placed in an aqueous electrolyte and exposed to sunlight (Figure 7.5a). One can combine it with the energy diagram (Figure 7.5b), which is similar to that shown in Figure 7.3 with the physical scale, to result in a combined picture convenient to use while considering various semiconductor-based systems common in artificial photosynthesis (Figure 7.5c).

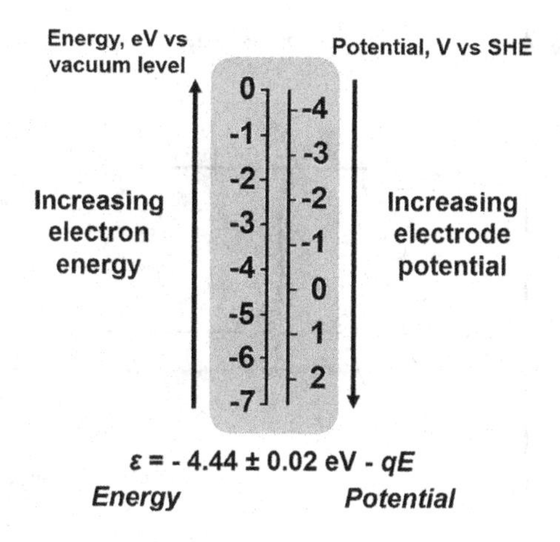

FIGURE 7.4 Absolute energy vs. electrochemical scales.

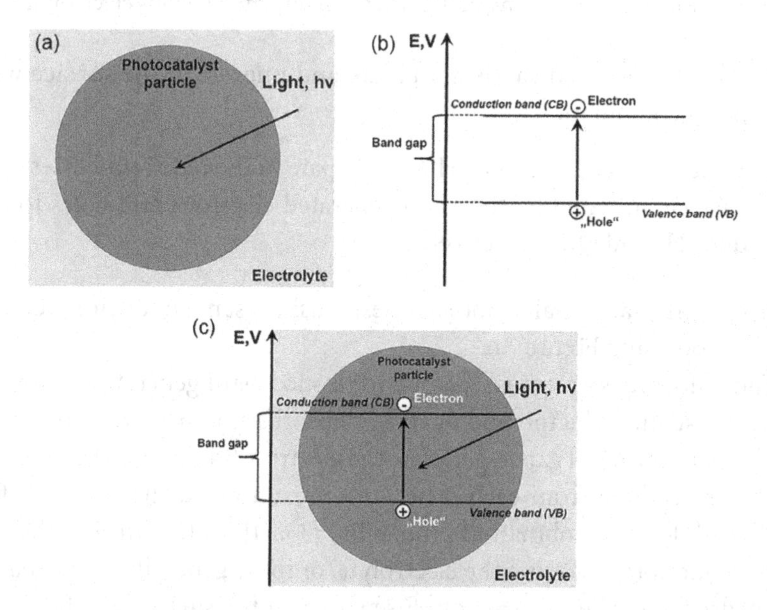

FIGURE 7.5 Construction of a model used in the field of artificial photosynthesis. (a) Schematic drawing of a photocatalyst particle in an aqueous electrolyte under illumination. (b) Schematic energy diagram. (c) Combined picture.

7.3 ONE-STEP PHOTOEXCITATION SYSTEMS

Let us now name factors affecting the activity of photocatalytic materials. The main factors are related to the following three steps:

1. The photocatalyst absorbs photons of higher energy than the band gap energy of the material and generates photoexcited electron-hole pairs in bulk

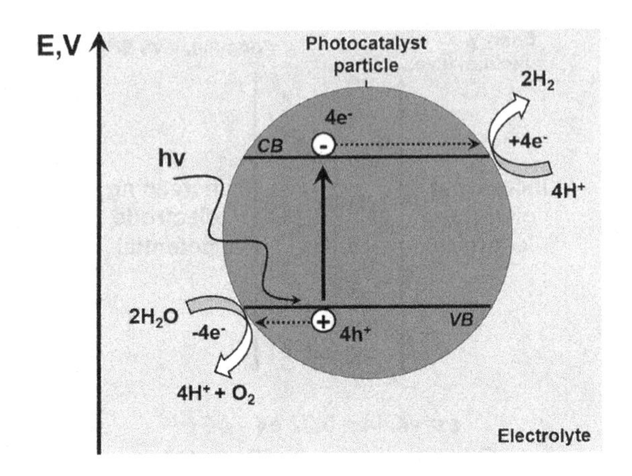

FIGURE 7.6 Conceptual scheme of artificial photosynthesis to produce hydrogen fuel using semiconducting materials, sunlight, and aqueous electrolytes.

2. The photoexcited carriers separate and migrate to the surface without recombination
3. Water molecules (and/or H^+) participate in the redox processes at the surface, initiated by the photogenerated electrons and holes to produce H_2 and O_2, respectively

Conceptually, artificial photosynthesis using semiconductors can be understood using Figure 7.6.

The following steps must happen for a successful generation of hydrogen. The semiconductor photocatalyst absorbs photons with an energy higher than the band gap to generate the electron-hole pairs. The holes and electrons separate from each other and move toward the surface, effectively with low recombination probability [33,34]. At the surface, the electrons reduce protons from the electrolyte, or more generally, participate in the hydrogen evolution reaction. Elsewhere on the surface, the holes participate in the oxidation reaction resulting in gaseous oxygen. Obviously, these processes strongly depend on the structural and electronic properties of the photocatalyst.

What is the minimal energy of the absorbed photon to split water, and are there enough of such photons in the incoming sunlight? One can look at Figure 7.7, where the solar spectrum is presented. From previous considerations, it is known that the minimal energy of an incoming photon to split water is ~1.23 eV, corresponding to a wavelength of ~1000 nm, in the near-infrared region. Already, this excludes a considerable part of the solar spectrum. In addition to the overpotential for the hydrogen and

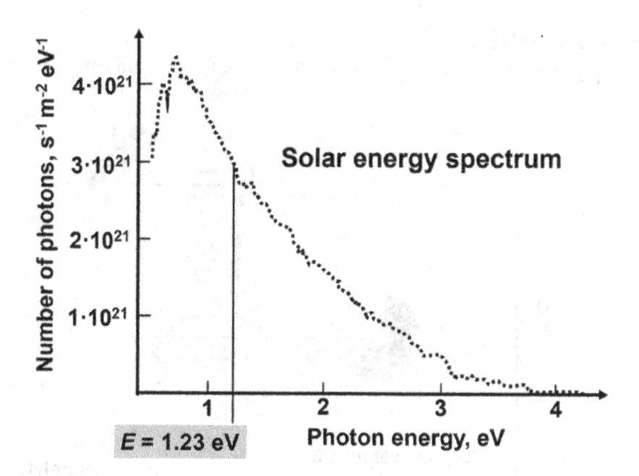

FIGURE 7.7 Approximate solar energy spectrum. Theoretical energy to split water (1.23 eV) is also indicated. Due to the considerable overpotentials for water splitting, one should mainly consider photons with energy larger than ca 2 eV.

oxygen evolution reactions, one needs to account for the noticeable losses due to the complications in the catalysis of these processes. Therefore, the energy of the photons, which can be utilized for artificial photosynthesis, must be higher than *ca* 1.8–2 eV. As one can see from Figure 7.7, there are still plenty of photons with such energy reaching the planet's surface [35].

However, the energy of incoming photons is only a part of the story. The energy of the photogenerated electrons and holes must match the potentials where the hydrogen evolution reaction, HER, and oxygen evolution reaction, OER, occur. In other words, to achieve the overall water splitting, the bottom of the conduction band must be more negative than the potential for the H_2 evolution reaction (which is −0.41 V vs. SHE at pH=7, or 0.0 V vs. RHE scale). At the same time, the top of the valence band must be more positive than the potential for the oxygen evolution reaction (0.82 V vs. SHE at pH=7 or 1.23 V vs. RHE scale).

This situation is reflected in more detail in Figure 7.8 for an ideal case. However, the right valence and conduction bands' position is still not enough for efficient water splitting, as most photocatalysts are not particularly good electrocatalysts for oxygen evolution and hydrogen evolution reactions. To reduce losses caused by the respective overpotentials and enable good selectivity, the involvement of the so-called cocatalysts is highly desirable [36–38]. These cocatalysts can be chosen from the catalytic materials known in electrocatalysis. For instance, one can use nickel oxides to accelerate the oxygen evolution reaction and metallic nickel for the HER in alkaline media. Likewise, metallic platinum can be the catalyst of choice for the hydrogen evolution and IrO_x for the OER in acidic media.

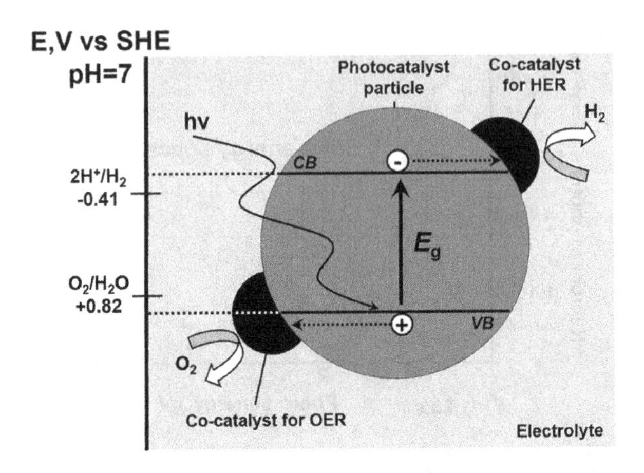

FIGURE 7.8 Schematic energy diagram for the photocatalytic water splitting for a one-step photoexcitation system. CB stands for the conduction band; VB – valence band; E_g – band gap. Note the relative positions of the valence and conduction bands versus the equilibrium potentials of the hydrogen and oxygen evolution reactions. Cocatalysts accelerating the OER and HER should be used in order to decrease the overpotentials for these reactions. (Adapted from [39].)

However, now the question arises, where exactly at the surface should one immobilize the cocatalysts? Figure 7.8 shows the concept, but such a model does not give instructions on where to locate the cocatalysts.

From the straightforward approach, it is logical to put cocatalysts at the surface regions where either electrons or holes "go out". In other words, the material catalyzing the oxygen evolution reaction should be placed close to the surface sites, where holes "go out". Consequently, the cocatalysts promoting the hydrogen evolution should be immobilized at the places where the probability of meeting the photogenerated electron is the highest. However, is it known if electrons and holes indeed appear at preferential specific locations at the surface? Until recently, there was a dominating hypothesis that both of them appear at the surface and participate in the redox catalytic reactions randomly. Therefore, researchers tried to distribute cocatalysts evenly at the surface of photocatalysts. However, recently, it has been shown that it is in many cases not true, and there are preferential places for the generated electrons and holes at the surface to initiate the reduction and oxidation reactions [40–42]. For instance, under illumination, the photogenerated electrons in photocatalytic $BiVO_4$ tend to accumulate at the [010] plane, while the holes preferentially move toward the [110] facets (Figure 7.9).

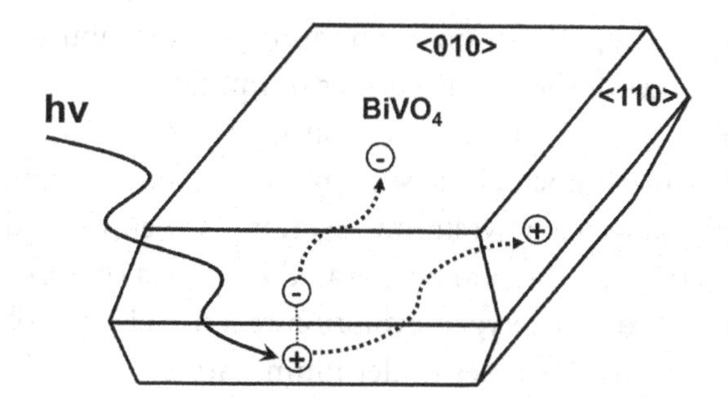

FIGURE 7.9 The spatial separation of the photogenerated electron-hole pair at the semiconducting $BiVO_4$ photocatalyst. The electrons preferentially go toward the [010] planes while the holes move to the [110] facets. (Adapted from [40].)

It is possible to visualize the areas at the surface, which are reached by the photogenerated electrons. For that, one can reduce, for example, dissolved platinum precursors under illumination using those electrons. Figure 7.10 illustrates that the reduction process indeed occurs only at specific facets resulting in metallic platinum particles at the surface. Suppose the surrounding electrolyte is afterward replaced with another one containing a dissolved Mn-precursor, and again the system is exposed to the sunlight. In that case, it is possible to initiate deposition of Mn oxides at the facets, where the holes "go out", as the holes oxidize Mn^{2+} ions in the electrolyte forming a deposit. As one can see from Figure 7.10, the resulting MnO_x deposit is located at specific facets of the photocatalyst crystal.

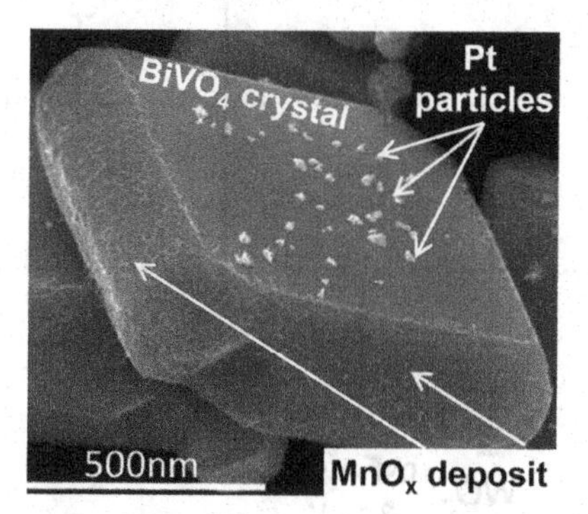

FIGURE 7.10 Scanning electron microscopy image of a $BiVO_4$ photocatalyst with deposited metallic Pt at the facets where photogenerated electrons initiate the reduction of a Pt-precursor from the solution. At the same time, the holes initiate local deposition of Mn oxide at completely different surface planes. (Adapted from [40].)

This confirms the hypothesis on the nonuniform distribution of the photogenerated charges at the surface under illumination.

Remarkably, using the same deposition approach, one can design potentially very efficient photocatalytic systems [43], as Pt, being one of the best catalysts for the hydrogen evolution reaction, is then located precisely at the right place to meet the photogenerated electrons (Figure 7.10). MnO_x, a suitable catalyst for the oxygen evolution reaction, is immobilized at the regions enriched with the holes under illumination.

Unfortunately, there is a certain lack of suitable photocatalysts that simultaneously meet the following three requirements:

1. The band gap is narrower than *ca* 3 eV but wider than *ca* 2 eV
2. Band edge potentials are appropriate for the overall water splitting
3. Good stability under the reaction conditions

Figure 7.11 schematically shows the energy diagrams of some well-known semiconducting materials with respect to the equilibrium potentials of the HER and OER.

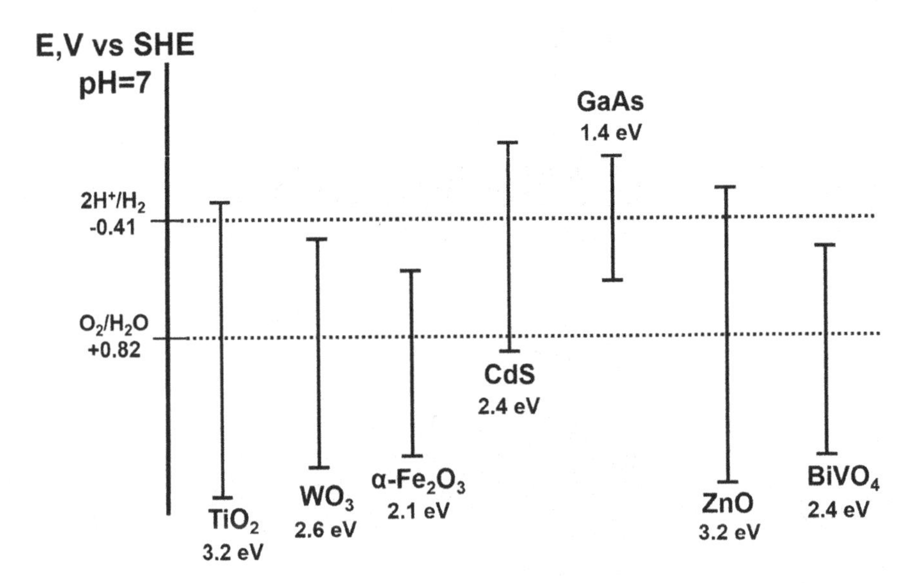

FIGURE 7.11 Approximate positions of the conduction and valence bands versus the equilibrium potentials for the hydrogen and oxygen evolution reactions in aqueous electrolytes at pH = 7. Band gaps are also given. (Data for $BiVO_4$ are from [44].)

As one can see from the Figure, only a few compounds can be considered for one-photon water splitting. From Figure 7.11, one can pick TiO_2, CdS, and ZnO. At the first glance, CdS could be the best candidate for artificial photosynthesis [45]. The band gap (*ca* 2.4 eV) is just slightly wider than ideally required, and the positions of the valence and conduction bands are appropriate to initiate both hydrogen and oxygen evolution reactions. Unfortunately, stability issues restrict so far its use as a photocatalyst. It turned out that metal chalcogenides, without external protection by, e.g., thin layers of some oxides, are often not stable in water oxidation reaction to form O_2. It is because the S^{2-} and Se^{2-} in the photocatalysts are more prone to oxidation than water, causing the CdS or CdSe catalysts themselves to be oxidized and degraded (the so-called photocorrosion). On the other hand, TiO_2 and ZnO have a pretty wide band gap (*ca* 3.2 eV), enabling only a tiny part of the incoming photons to be absorbed (see Figure 7.7). Nevertheless, those oxides, especially titania, appeared to be very stable under illumination and HER/OER conditions. Therefore, TiO_2 and doped titania are widely considered as the state-of-the-art materials in the discussed applications [46,47].

Titania surfaces are not good catalysts for neither the hydrogen evolution nor for the oxygen evolution reactions. Therefore, the photocatalytic systems based on TiO_2 need cocatalysts. However, even with such a modification, the resulting systems' productivity in the sense of the amounts of generated hydrogen is pretty low due to the intrinsic drawback of titania – a too wide band gap.

Addressing the challenge with a relatively poor choice of photocatalytic materials for the one-photon artificial photosynthesis, a number of research groups worldwide try to perform a thorough theoretical screening of new possible candidates. One of the first examples of computational high-throughput screening using density functional theory calculations is dated 2012, when several promising materials with the perovskite structure were theoretically identified (Figure 7.12) [48].

Further requirements for the photoactive systems include the following. The photocatalysts should be highly crystalline. In general, high crystallinity positively affects the system's performance since the density of defects, acting as recombination centers between photogenerated charge carriers, decreases with increasing crystallinity. One can also expect higher photocatalytic activity by reducing the particle size of a photocatalyst since the diffusion length for the photogenerated electron-hole pairs can be shortened. It is essential to design both the bulk and surface properties of

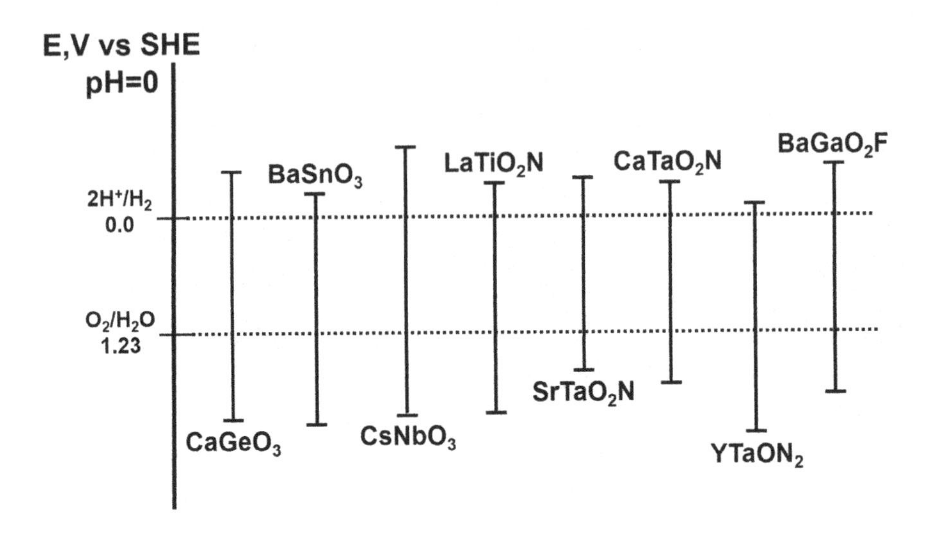

FIGURE 7.12 Some theoretically identified candidates with the cubic perovskite structure for one-photon photocatalytic water splitting. (The data are from [48].)

the material simultaneously and carefully to obtain a high activity for the water-splitting reaction.

While there are many indications that the one-photon water splitting should be possible with reasonable efficiency, reproducible and stable photocatalytic systems for a visible-light-driven one photon scheme had not been realized by now. However, one-step water splitting by visible light had been once described as one of the "holy grail" of materials science. Nevertheless, it is possible to alternatively suggest a wider choice of materials if one designs systems with the so-called two-photon water splitting, which is the topic of the next section.

7.4 TWO-STEP PHOTOEXCITATION SYSTEMS

Consider two particles made of different semiconducting materials (Figure 7.13). Imagine also that each semiconductor has a moderate band gap. Let us also assume that the position of the valence band of one semiconductor is appropriate to initiate the oxygen evolution reaction at its surface. At the same time, the position of the conduction band is too positive to start the HER. For the second semiconductor, the situation is different: The bottom of the conduction band is located more negative than the equilibrium potentials for the hydrogen evolution, but the position of the valence band prohibits the OER (see Figure 7.13). Assume also appropriate "acceptor" and "donor" electroactive species (redox mediators, acting as a

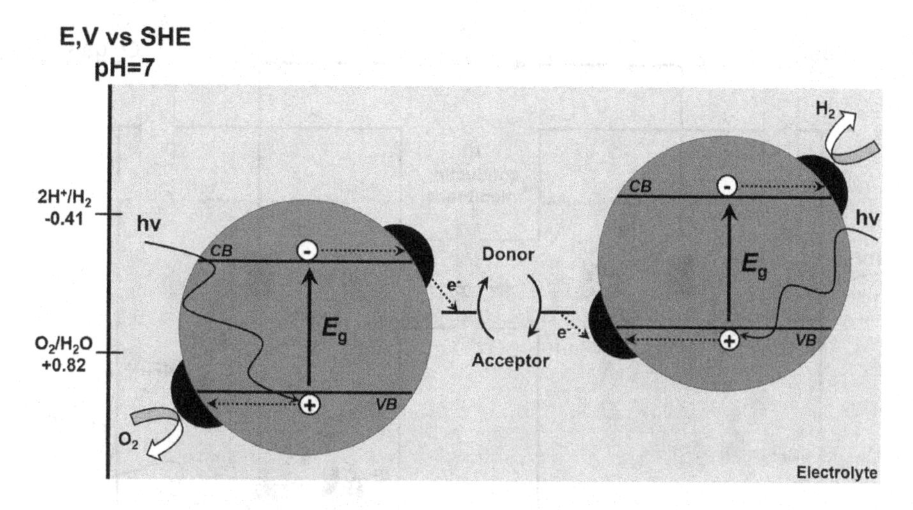

FIGURE 7.13 Schematic energy diagram of photocatalytic water splitting for a two-step photoexcitation system. CB – conduction band, VB – valence band, E_g – band gap; donor/acceptor indicate electron-donating and accepting species, respectively.

"charge shuttle") with the redox potential positioned in the energy scale between the conduction band of the first semiconductor and the valence band of the second one. The essential property of such an acceptor/donor system should be the very fast rate of electron transfer. This is possible if some outer-sphere reactions, not catalytic ones, are selected.

Take a look at Figure 7.13. Imagine the particle on the left side absorbs the first photon, resulting in an electron/hole pair formation. The hole can participate in the OER while the photogenerated electron from the conduction band reduces the acceptor species to result in the donor one. Another photon is absorbed by the second semiconductor particle (on the right in Figure 7.13), also generating an electron/hole pair. The corresponding hole is then used to oxidize the donor to regenerate the acceptor species while the photogenerated electron in the corresponding conduction band participates in the hydrogen evolution reaction. Thus, both the hydrogen and oxygen evolution reactions are enabled even though the individual photocatalysts are not optimal to initiate both reactions.

In the following, it is important to design a more realistic system so that the hydrogen and oxygen evolution reactions are spatially separated to collect pure H_2 fuel. Figure 7.14 shows one example of such a design with photocatalysts, cocatalysts, and redox mediator pairs. In the right part of Figure 7.14, the scheme shows the oxygen evolution reaction under

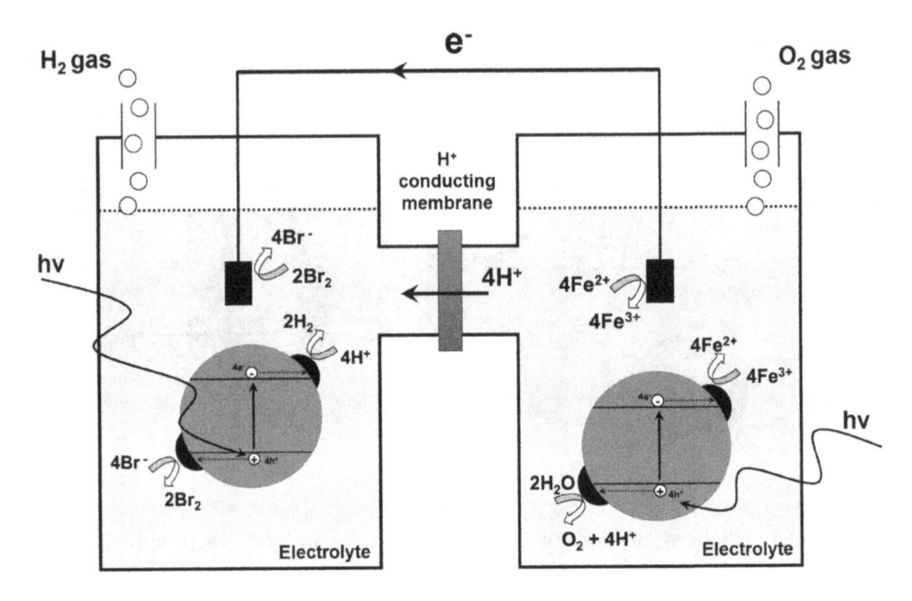

FIGURE 7.14 Schematic illustration of a model two-photon water-splitting cell using a photocatalyst in the presence of $Br_2/2Br^-$ and Fe^{3+}/Fe^{2+} redox mediators. Besides the catalytic HER and OER, Fe^{2+} ions produced as a result of reduction at the photocatalyst (right part) react spontaneously with Br_2, which is, in turn, the oxidation product at the photocatalyst in the left part, using an electron conductor connecting two cell compartments, thereby regenerating Fe^{3+} and Br^-. (Adapted from [39].)

illumination with the participation of photogenerated holes at an immobilized cocatalyst (e.g., Ir oxide). At the same time, the photogenerated electrons quickly reduce Fe^{3+} to Fe^{2+}.

The schematics on the left part of Figure 7.14 illustrate the hydrogen evolution reaction using photogenerated electrons at a cocatalyst (e.g., Pt), while the generated holes oxidize Br^- ions to give Br_2. Finally, spontaneous spatially separated redox process organized through an electron conductor regenerates Br^- and Fe^{3+}: $Br_2+2Fe^{2+} \rightarrow 2Br^-+2Fe^{3+}$. It should also be noted that a proton-conducting membrane separates the two compartments but allows the transfer of the protons generated due to OER to the left compartment, as shown in Figure 7.14.

Here, it is noteworthy to discuss the choice of the cocatalysts. As discussed previously, the most challenging reaction to catalyze in the systems described in this chapter is the oxygen evolution reaction. The choice of active catalysts is quite limited, and for the artificial leaf concept, one needs a very affordable catalyst candidate. In principle, one might consider

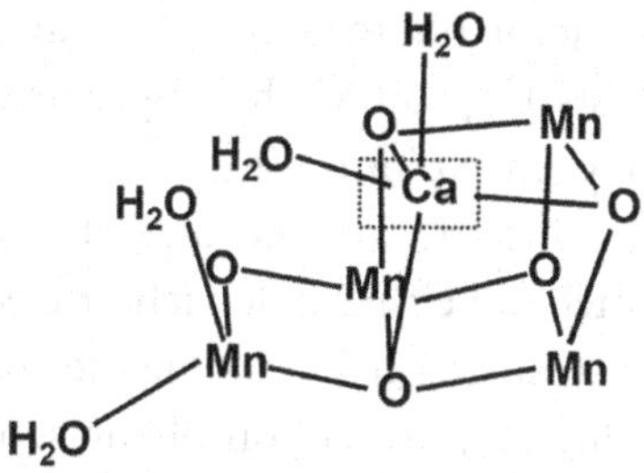

FIGURE 7.15 Mn_4O_5Ca cluster of Photosystem II: a schematic structural formula.

starting with the immobilization of natural objects, which are constituents of the bio-systems. For example, one can try to immobilize the water oxidizing complex of Photosystem II directly. The latter is the only system to catalyze water oxidation in nature, and the water oxidation center is a Mn_4O_5Ca cluster (see Figure 7.15) accommodated in a protein environment of the Photosystem II that controls the reaction coordinates, proton movement, and water access. However, most of them were tested not to be stable enough under harsh conditions of the OER for industrial applications. Nevertheless, one can use it as a starting point, as a sort of structural motif to improve the performance in the future.

7.5 SUMMARY AND CONCLUSIONS

Photoinduced water splitting has recently attracted significant attention. It would probably be an ideal approach to mimic the natural processes, harvest the energy of light, and produce fuels such as hydrogen. The most convenient approach to artificial photosynthesis is using semiconductor materials with appropriate band gaps (a bit wider than 2 eV) for one-photon water splitting. However, only a few materials are known for that with appropriate band gaps and positions of the valence and conduction bands, which are not stable under illumination. All the known artificial systems utilizing the one-photon scheme suffer either from photocorrosion or low efficiency. Relatively wide band gap materials like TiO_2 are considered to be the state of the art. Recent attempts to address the challenge include high throughput theoretical and experimental screening.

An alternative approach to enable a wider choice of semiconductors is to use the two-photon schemes with appropriate intermediate reversible redox mediators. Several promising designs have been proposed. However, further research and development are at the moment necessary for the commercialization of these systems.

The use of semiconductor photocatalysts for artificial photosynthesis often requires cocatalysts to promote hydrogen evolution (Pt, nickel or their alloys, etc.) and oxygen evolution (e.g., nickel oxides, IrO_2) reactions. In order to increase the efficiency, the cocatalysts should be immobilized at the surface areas, which are either hole-rich or electron-rich under illumination. Other recent strategies to overcome the challenges are the rational design of semiconductors, protection of the semiconductor surfaces, use of molecular complexes, and immobilization of objects of nature.

7.6 QUESTIONS

1. Explain the concept of "artificial leaves".
2. What are the main approaches to artificial photosynthesis?
3. How can one use semiconductor materials in water splitting without an external bias?
4. Define photocatalysts and cocatalysts.
5. What are the main factors affecting the activity of photocatalytic materials?
6. Explain the idea of the two-step photoexcitation systems for water splitting.
7. Analyze promises and challenges concerning the "artificial leaf" approach.

REFERENCES

1. Lubitz. W.; Reijerse, E.J.; Messinger, J. 2008. Solar water-splitting into H_2 and O_2: Design principles of photosystem II and hydrogenases. *Energy & Environmental Science* 1:15–31.
2. Dalle, K.E.; Warnan, J.; Leung, J.J.; Reuillard, B.; Karmel, I.S.; Reisner, E. 2019. Electro- and solar-driven fuel synthesis with first row transition metal complexes. *Chemical Reviews* 119:2752–2875.
3. Scholes, G.D.; Fleming, G.R.; Olaya-Castro, A.; van Grondelle, R. 2011. Lessons from nature about solar light harvesting. *Nature Chemistry* 3:763–774.
4. Barber, J. 2006. Photosystem II: An enzyme of global significance. *Biochemical Society Transactions* 34:619–631.
5. Vinyard, D.J.; Ananyev, G.M.; Dismukes, G.C. 2013. Photosystem II: The reaction center of oxygenic photosynthesis. *Annual Review of Biochemistry* 82:577–606.

6. Kim, J.H.; Hansora, D.; Sharma, P.; Jang, J.W.; Lee, J.S. 2019. Toward practical solar hydrogen production – an artificial photosynthetic leaf-to-farm challenge. *Chemical Society Reviews* 48:1908–1971.
7. Ismail, A.A.; Bahnemann, D.W. 2014. Photochemical splitting of water for hydrogen production by photocatalysis: A review. *Solar Energy Materials and Solar Cells* 128:85–101.
8. Li, X.B.; Xin, Z.K.; Xia, S.G.; Gao, X.Y.; Tung, C.H.; Wu, L.Z. 2020. Semiconductor nanocrystals for small molecule activation via artificial photosynthesis. *Chemical Society Reviews* 49:9028–9056.
9. Gust, D.; Moore, T.A.; Moore, A.L. 2012. Realizing artificial photosynthesis. *Faraday Discussions* 155:9–26.
10. Bard, A.J.; Fox, M.A. 1995. Artificial photosynthesis: Solar splitting of water to hydrogen and oxygen. *Accounts of Chemical Research* 28:141–145.
11. Copeland, A.W.; Black, O.D.; Garrett, A.B. 1942. The photovoltaic effect. *Chemical Reviews* 31:177–226.
12. Bockris, J.O'M.; Handley, L. 1978. On photo fuel. *Energy Conversion* 18:1–8.
13. House, R.L.; Iha, N.Y.M.; Coppo, R.L.; Alibabaei, L.; Sherman, B.D.; Kang, P.; Brennaman, M.K.; Hoertz, P.G.; Meyer, T.J. 2015. Artificial photosynthesis: Where are we now? Where can we go? *Journal of Photochemistry and Photobiology C: Photochemistry Reviews* 25:32–45.
14. Tournet, J.; Lee, Y.; Karuturi, S.K.; Tan, H.H.; Jagadish, C. 2020. III–V semiconductor materials for solar hydrogen production: Status and prospects. *ACS Energy Letters* 5:611–622.
15. Wu, H.; Tan, H.L.; Toe, C.Y.; Scott, J.; Wang, L.; Amal, R.; Ng, Y.H. 2020. Photocatalytic and photoelectrochemical systems: Similarities and differences. *Advanced Materials* 32:1904717.
16. Fang, X.; Kalathil, S.; Reisner, E. 2020. Semi-biological approaches to solar-to-chemical conversion. *Chemical Society Reviews* 49:4926–4952.
17. Fukuzumi, S.; Lee, Y.M.; Nam, W. 2020. Bioinspired artificial photosynthesis systems. *Tetrahedron* 76:131024.
18. Gaut, N.J.; Adamala, K.P. 2020. Toward artificial photosynthesis. *Science* 368:587–588.
19. Huang, J.; Galluccia, J.C.; Turro, C. 2020. Panchromatic dirhodium photocatalysts for dihydrogen generation with red light. *Chemical Science* 11:9775–9783.
20. Llansola-Portoles, M.J.; Palacios, R.E.; Gust, D.; Moore, T.A.; Moore, A.L. 2015. Artificial photosynthesis: From molecular to hybrid nanoconstructs. In: *From Molecules to Materials*. Editors: Rozhkova, E.; Ariga K. Springer: Cham.
21. Zhang, B.; Sun, L. 2019. Artificial photosynthesis: Opportunities and challenges of molecular catalysts. *Chemical Society Reviews* 48:2216–2264.
22. Liu, C.; van den Bos, D.; den Hartog, B.; van der Meij, D.; Ramakrishnan, A.; Bonnet, S. 2021. Ligand controls the activity of light-driven water oxidation catalyzed by nickel(II) porphyrin complexes in neutral homogeneous aqueous solutions. *Angewandte Chemie International Edition* 133:13575–13581.

23. Frischmann, P.D.; Mahata, K.; Würthner, F. 2013. Powering the future of molecular artificial photosynthesis with light-harvesting metallosupramolecular dye assemblies. *Chemical Society Reviews* 42:1847–1870.
24. Moss, B.; Babacan, O.; Kafizas, A.; Hankin, A. 2021. A Review of inorganic photoelectrode developments and reactor scale-up challenges for solar hydrogen production. *Advanced Energy Materials* 11:2003286.
25. Landman, A.; Dotan, H.; Shter, G.E.; Wullenkord, M.; Houaijia, A.; Maljusch, A.; Grader, G.S.; Rothschild, A. 2017. Photoelectrochemical water splitting in separateoxygen and hydrogen cells. *Nature Materials* 16:646–651.
26. Kim, T.W.; Choi, K.S. 2014. Nanoporous $BiVO_4$ photoanodes with dual-layer oxygen evolution catalysts for solar water splitting. *Science* 343:990–994.
27. Hisatomi, T.; Takanabe, K.; Domen, K. 2015. Photocatalytic water-splitting reaction from catalytic and kinetic perspectives. *Catalysis Letters* 145:95–108.
28. Maeda, K. 2011. Photocatalytic water splitting using semiconductor particles: History and recent developments. *Journal of Photochemistry and Photobiology C: Photochemistry Reviews* 12:237–268.
29. Li, L.; Yan, J.; Wang, T.; Zhao, Z.J.; Zhang, J.; Gong, J. 2015. Sub-10 nm rutile titanium dioxide nanoparticles for efficient visible-light-driven photocatalytic hydrogen production. *Nature Communications* 6:5881.
30. Hisatomi, T.; Domen, K. 2019. Reaction systems for solar hydrogen production via water splitting with particulate semiconductor photocatalysts. *Nature Catalysis* 2:387–399.
31. Kisch, H. 2015. *Semiconductor Photocatalysis: Principles and Applications.* Wiley VCH: Weinheim.
32. Trasatti, S. 1986. The absolute electrode potential: An explanatory note. *Pure & Applied Chemistry* 58:955–966.
33. Cowana, A.J.; Durrant, J.R. 2013. Long-lived charge separated states in nanostructured semiconductor photoelectrodes for the production of solar fuels. *Chemical Society Reviews* 42:2281–2293.
34. Osterloh, F.E. 2013. Inorganic nanostructures for photoelectrochemical and photocatalytic water splitting. *Chemical Society Reviews* 42:2294–2320.
35. Bak, T.; Nowotny, J.; Rekas, M.; Sorrell, C.C. 2002. Photo-electrochemical hydrogen generation from water using solar energy. Materials-related aspects. *International Journal of Hydrogen Energy* 27:991–1022.
36. Busser, G.W.; Mei, B.; Weide, P.; Vesborg, P.C.K.; Stührenberg, K.; Bauer, M.; Huang, X.; Willinger, M.-G.; Chorkendorff, I.; Schlögl, R.; Muhler, M. 2015. Cocatalyst designing: a regenerable molybdenum-containing ternary cocatalyst system for efficient photocatalytic water splitting. *ACS Catalysis* 5:5530–5539.
37. Li, X.; Yu, J.; Jaroniec, M.; Chen, X. 2019. Cocatalysts for selective photoreduction of CO_2 into solar fuels. *Chemical Reviews* 119:3962–4179.
38. Ye, S.; Ding, C.; Liu, M.; Wang, A.; Huang, Q.; Li, C. 2019. Water oxidation catalysts for artificial photosynthesis. *Advanced Materials* 31:1902069.
39. Maeda, K. 2013. Z-Scheme water splitting using two different semiconductor photocatalysts. *ACS Catalysis* 3:1486–1503.

40. Li, R; Han, H.; Zhang, F.; Wang, D.; Li, C. 2014. Highly efficient photocatalysts constructed by rational assembly of dual-cocatalysts separately on different facets of $BiVO_4$. *Energy & Environmental Science* 7:1369–1376.
41. Mu, L.; Zhao, Y.; Li, A.; Wang, S.; Wang, Z.; Yang, J.; Wang, Y.; Liu, T.; Chen, R.; Zhu, J.; Fan, F.; Li, R.; Li, C. 2016. Enhancing charge separation on high symmetry $SrTiO_3$ exposed with anisotropic facets for photocatalytic water splitting. *Energy & Environmental Science* 9:2463–2469.
42. Zhu, J.; Fan, F.; Chen, R; An, H.; Feng, Z.; Li, C. 2015. Direct imaging of highly anisotropic photogenerated charge separations on different facets of a single $BiVO_4$ photocatalyst. *Angewandte Chemie International Edition* 54:9111–9114.
43. Adler, C.; Mitoraj, D.; Krivtsov, I.; Beranek, R. 2020. On the importance of catalysis in photocatalysis: Triggering of photocatalysis at well-defined anatase TiO_2 crystals through facet-specific deposition of oxygen reduction cocatalyst. *The Journal of Chemical Physics* 152: 244702.
44. Li, Y.L.; Liu, Y.; Hao, Y.J.; Wang, X.J.; Liu, R.H.; Li, F.T. 2020. Fabrication of core-shell $BiVO_4$@Fe_2O_3 heterojunctions for realizing photocatalytic hydrogen evolution *via* conduction band elevation. *Materials & Design* 187:108379.
45. Wolff, C.M.; Frischmann, P.D.; Schulze, M.; Bohn, B.J.; Wein, R.; Livadas, P.; Carlson, M.T.; Jäckel, F.; Feldmann, J.; Würthner, F.; Stolarczyk, J.K. 2018. All-in-one visible-light-driven water splitting by combining nanoparticulate and molecular co-catalysts on CdS nanorods. *Nature Energy* 3:862–869.
46. Li, Y.; Peng, Y.K. Hu, L.; Zheng, J.; Prabhakaran, D.; Wu, S.; Puchtler, T.J.; Li, M.; Wong, K.Y.; Taylor, R.A.; Tsang S.C.E. 2019. Photocatalytic water splitting by N-TiO_2 on MgO (111) with exceptional quantum efficiencies at elevated temperatures. *Nature Communications* 10:4421.
47. Asahi, R.; Morikawa, T.; Irie, H.; Ohwaki, T. 2014. Nitrogen-doped titanium dioxide as visible-light-sensitive photocatalyst: Designs, developments, and prospects. *Chemical Reviews* 114:9824–9852.
48. Castelli, I.E.; Landis, D.D.; Thygesen, K.S.; Dahl, S.; Chorkendorff, I.; Jaramillo, T.F.; Jacobsen, K.W. 2012. New cubic perovskites for one- and two-photon water splitting using the computational materials repository. *Energy & Environmental Science* 5:9034–9043.

CHAPTER 8

Materials for Solar Cell Applications

8.1 MOTIVATION AND HISTORICAL DEVELOPMENT

The potential of solar energy is roughly assessed to be 1.76×10^5 TW striking our planet, taking into account the degree to which the Earth reflects light from the sun. Human-related energy consumption in 2019 was ~18 TW. Obviously, the development of means and technologies to accumulate and convert solar energy is extremely attractive and viable. One way to do so is to use the "artificial leaf" approach to generate hydrogen fuel, as discussed in the previous chapter. An alternative way is to use photovoltaic (PV) devices such as solar cells to provide electricity.

The PV effect, which is nowadays used in common solar cells and can be arbitrarily designated as the beginning of the development in the field, was discovered by Edmond Becquerel in the 19th century [1] while performing experiments with solid electrodes in contact with electrolytes. He observed a direct current in the external circuit under the illumination of electrodes. The "best results" in such experiments were obtained with blue or ultraviolet light and when electrodes were coated with light-sensitive materials, such as AgCl or AgBr.

One can admit that thin-film-like PV devices using Se were suggested already in the 19th century by Adams and Day (1876) and Fritts (1883), among others. The explanation of the origin related to the PV effect was started by A. Einstein much later, in 1905, by first explaining the photoelectric effect [2]. One of the first silicon-based PV cells was proposed by

DOI: 10.1201/9781003025498-8

Ohl in 1941 [3,4]. Further milestones included Si cells for ~1500 USD/W with 2% efficiency presented by Hoffman Electronics in 1955 and 10% efficient cells presented by the same firm in 1959. In 1963, Sharp Co (Japan) produced the first commercial Si modules. By 2020, more than 630 GW_p power (where "p" denotes the peak power) has been installed worldwide [5], which is nominally more than or at least similar to the cumulative nominal capacity of all nuclear reactors working nowadays taken together. According to the Fraunhofer Institute for Solar Energy Systems, the global annual production of solar cells in 2019 was equivalent to *ca* 140 GW_p, suggesting that the installed capacity will grow substantially in the near future. This is also supported by the fact that the price for solar cell modules has drastically decreased in the past decades (Figure 8.1).

However, the current price of the devices is essential to assess this technology as environmentally friendly and sustainable. For example, in the modern economy, one typically needs money to make more money. Similarly, it also takes energy to make or save energy. The concept of *energy payback* reflects this kind of idea. The energy payback estimates how long a state-of-the-art commercial PV system should operate to recover the energy and associated pollutions spent to make such a system. Depending on the geographical location, this time can range from 1.5 to 2 years in Northern Europe and approximately 1 year or even less for the south or equatorial regions worldwide. Considering a typical operational time of a solar panel of *ca* 20 years, even in the worst scenario, these kinds of solar cell systems can generate at least ten times the energy needed to manufacture it.

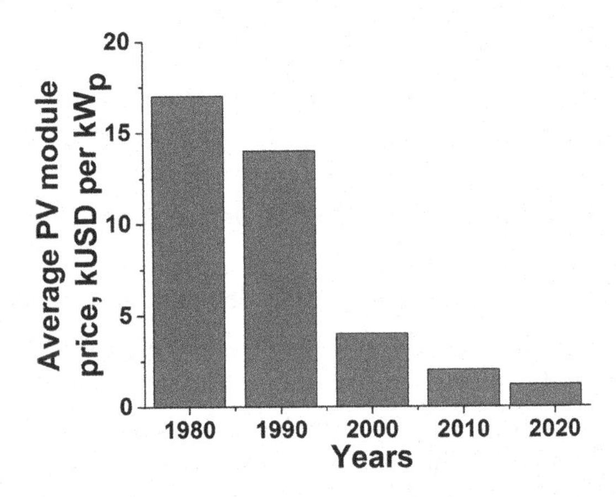

FIGURE 8.1 Average prices (in thousands United States Dollars, kUSD) for typical PV rooftop systems in 1980–2019. (The data are from [5,6].)

Moreover, the technology development within the last decade enabled reducing the required amount of pure Si from approximately 15g per W_p to *ca* 4g per W_p in 2020 [5]. The average price for electricity produced by PV systems is nowadays approximately the same or even lower than that produced using fossil fuels or nuclear powerplants. This is considering that initial subsidies for the PV technologies were less than half of the subsidies spent to support the nuclear energy's successful start [7]. Clearly, considering the net positive impact on the environment, solar cell technology experiences fast development even with certain recent acceleration [8].

8.2 SINGLE-JUNCTION SOLAR CELLS

In Chapter 7, only intrinsic undoped semiconductors were considered. However, to engineer standard solar cells, some doped semiconductors are necessary. Consider a pure Si material, which is a typical intrinsic semiconductor with a band gap of approximately 1.1 eV and the Fermi level lying in the middle between the valence and the conduction bands. Figure 8.2 schematically shows the crystal structure of Si with covalent bonds. Each silicon atom in such a structure forms four covalent bonds as Si has only four valent electrons. Consider replacing one of the silicon atoms from the crystal lattice with, for example, boron, B, which has only three valent electrons. If the amount of doping atoms is relatively small, the main structure remains while there will be a certain amount of defects in the covalent bonds (missing electrons, as shown in Figure 8.2). In this case, one can imagine that a positive charge is effectively introduced into the system. Therefore, this type of doped silicon is called p-type Si (p means positive).

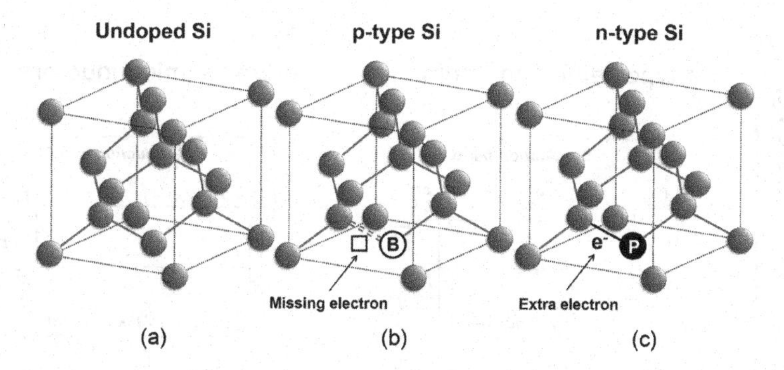

FIGURE 8.2 Doping of (a) pure Si with (b) boron, B, resulting in a defect in a covalent bond (missing electron) or (c) phosphorous, P, resulting in an "excessive" electron. Doping with B gives the p-type silicon, while doping with P gives the n-type silicon.

Imagine now that instead of boron, some atom with five valent electrons, for example, phosphorous, P, is introduced. In this case, one can arbitrarily say that an excessive electron and negative charge are introduced (see Figure 8.2). In this case, the doped Si is called n-Si (n means negative).

Let us consider how the energy diagrams should be modified compared to the cases discussed in Chapter 7 for the intrinsic semiconductors to describe the doped systems (Figure 8.3). Within the same framework, one can adjust the Fermi level to reflect the situation with both p-doping and n-doping. Recalling that the Fermi level corresponds to the density and the average energy of the quantum states of electrons, which are in the conduction and valence bands, n-doping will correspond to the shift of the Fermi level close to the conduction band. This new position will indicate an increased probability of finding an electron in the conduction band. Electrons in the conduction band are responsible for the increased electron conductivity, and therefore, the main charge carriers in the n-type semiconductors are electrons.

In contrast, the Fermi level should be shifted downwards in the case of p-type semiconductors. One can arbitrarily say that namely "holes" (defects in the covalent bonds) in the valence band present after the doping "can carry current". Therefore, they are formally the primary charge carriers in this kind of doped semiconductors.

Imagine one brings a p-type semiconductor in contact with a different medium, for instance, with another electronically conducting solid or an ionically conducting liquid having a dissimilar Fermi level (or equivalently with different electrochemical potential of electrons). Suppose

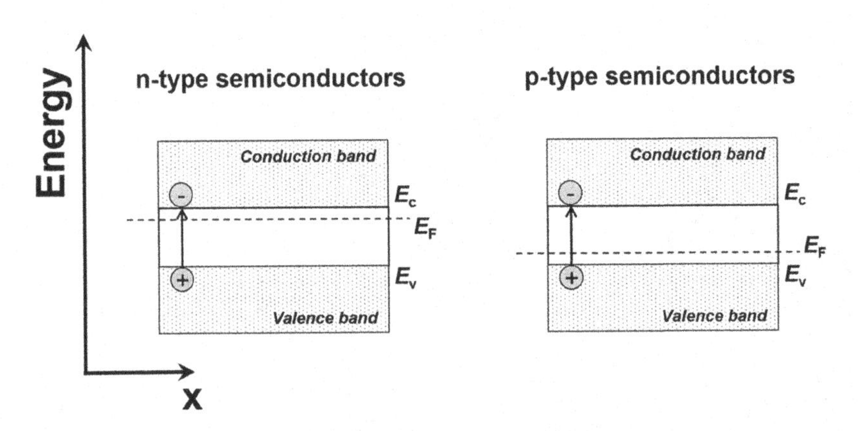

FIGURE 8.3 Schematic energy diagrams for n-type and p-type semiconductors. E_F stands for the Fermi level.

the E_F of the medium is located below the initial Fermi level of the semiconductor. After the contact, the Fermi levels should equilibrate. This means that the electrons from the semiconductors will reduce, e.g., the oxidized species in the electrolyte (see Figure 8.4a). In the case of metal electrodes, it would not considerably change the overall properties of the surface and subsurface regions of the electrode, as the number of charge careers in metals is very large. However, in semiconductors, such a redox process at the surface leads to the formation of a depletion layer, also called the space charge layer.

To designate this new situation in the energy diagram, one needs to introduce the *band bending* concept, as shown in Figure 8.4a. In the space

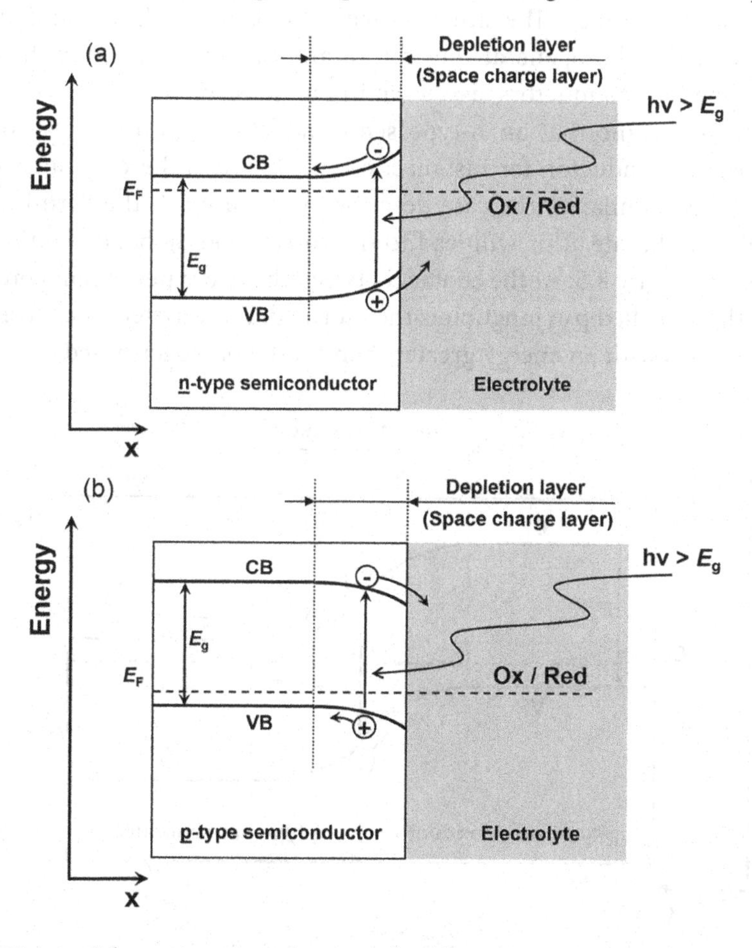

FIGURE 8.4 Schematic energy diagrams for (a) n-type and (b) p-type semiconductors in contact with an electrolyte containing a redox pair. CB and VB denote the conduction and valence bands, respectively.

charge layer, the number of electrons in the conduction band is reduced compared to the bulk, and the closer to the surface, the more the situation reminds the one for the undoped case. Therefore, using the formalism described before, the distance between the Fermi level and the conduction band increases if one moves toward the surface. If illumination with the photons having an energy greater than E_g is provided, it will generate an electron-hole pair. From the schematics shown in Figure 8.4a, it is evident that there will be a driving force for the generated photoelectron to move toward the bulk of the semiconductor. At the same time, there will be the driving force for the generated hole to move toward the surface of the electrode and participate in a redox reaction at the electrode/electrolyte interface. The situation with the band bending and driving forces for the photogenerated electrons and holes is inverted in the case of p-type semiconductors, as shown in Figure 8.4b.

Let us assume that an n-type semiconductor, e.g., n-type Si, and a p-type semiconductor, for instance, p-type Si, are in direct contact with each other. Similar to the case described in Figure 8.4, the Fermi levels should equilibrate. This will lead to the corresponding band bending, as shown in Figure 8.5. At the contact between the two types of semiconductors, the so-called p-n junction is then formed. If the system is exposed to the photons with an energy greater than E_g, the photogenerated electrons

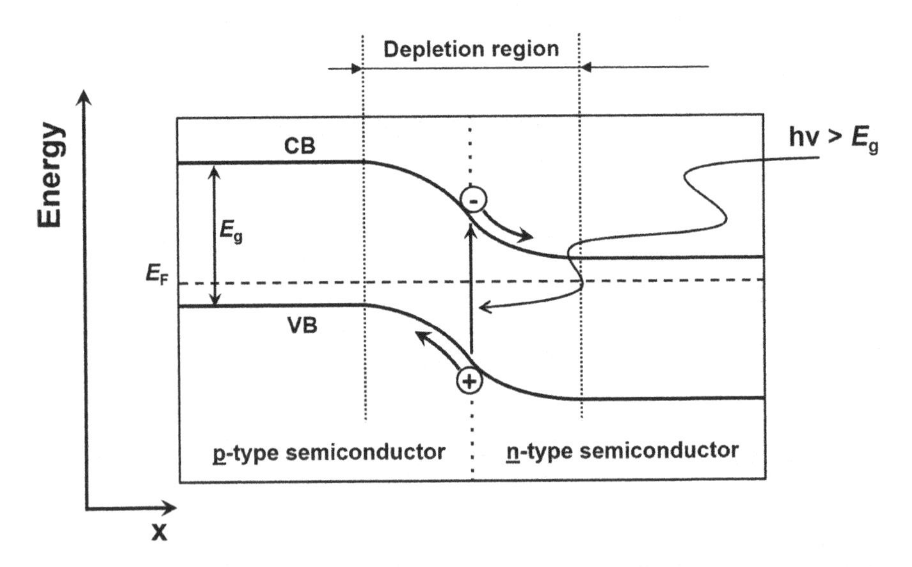

FIGURE 8.5 Schematic energy diagram for a p-n junction formed after the contact between p-type and n-type semiconductors.

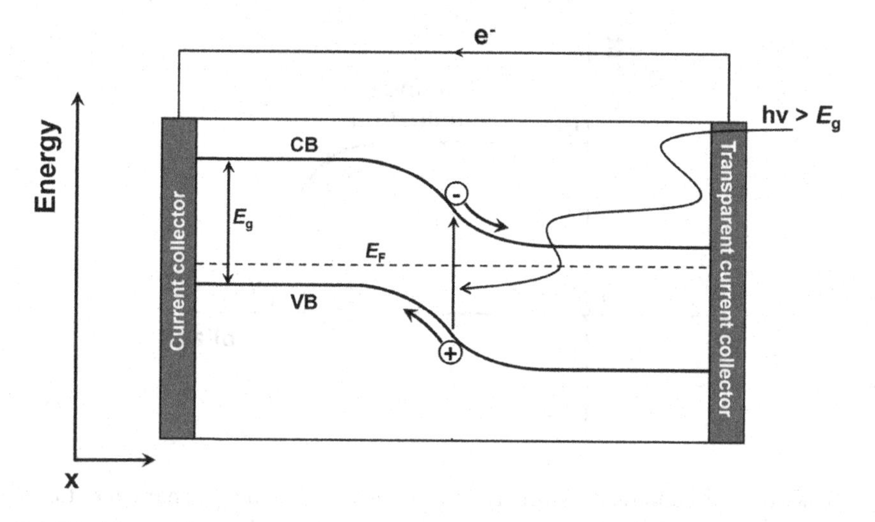

FIGURE 8.6 Schematics of an idealized solar cell under short circuit conditions.

and corresponding holes will tend to spatially separate at the p-n junction, like it is schematically explained in Figure 8.5.

There is now one step to constructing a device, a solar cell, which will convert the energy of the sunlight to electrical energy. It is necessary to introduce two electron-conducting current collectors, as shown in Figure 8.6, which describes a single p-n junction solar cell architecture. The critical issue is that at least one of the current collectors should be transparent to the visible light. More details on the transparent electron conductors are given in Chapter 9.

DEFINITION:

PV cells or solar cells can be defined as specific semiconductor diodes that convert (sun)light into direct electric current.

Usually, besides a transparent electron conductor, such as In-doped SnO_2, antireflection coatings, such as Si_3N_4 and metallic current collectors, commonly Al, are used. Each layer in the cell is stacked on top of the other.

In order to characterize the performance of solar cells, the current *versus* voltage curves are recorded under illumination. The schematic curve typical for solar cells recorded under illumination is shown in Figure 8.7. While without illumination, the I-V curve reveals a typical diode behavior (not shown), under illumination, the dependencies are changed, revealing several key points. Those are: (i) the current under the short circuit condition (this condition is depicted in Figure 8.6), I_{SC}, (ii) the open-circuit

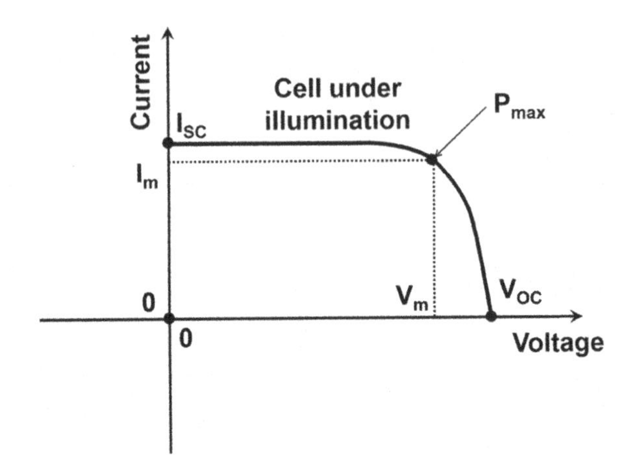

FIGURE 8.7 Schematic graph representing the current-voltage characteristics of a p-n junction under illumination. I_{SC} and V_{OC} stand for the short circuit current and the open-circuit voltage, respectively.

voltage, V_{OC}, and (iii) the maximum power, P_{max}, which is one of the important figures of merit, given as (see Figure 8.7):

$$P_{max} = V_m \cdot I_m$$

An essential parameter of a solar cell is its energy conversion efficiency. The latter is commonly defined as the percentage of power converted from sunlight to electrical energy under some standard test conditions. Probably the first quantitative estimation of the maximum theoretical efficiency of a single junction solar cell was performed in 1961 by W. Shockley and H. Queisser (Figure 8.8) [9]. For the estimation, the standard conditions and assumptions were set, as described below.

- Only one semiconductor material is used in the solar cell
- There is only one p-n junction per solar cell
- Not-concentrated sunlight is used. Practically, the common illumination conditions existing in California (USA) were selected for the first standardization
- All energy is converted to heat from photons having energy larger than the band gap. It is also assumed that the electron loses the additional energy when traveling toward the p-n junction

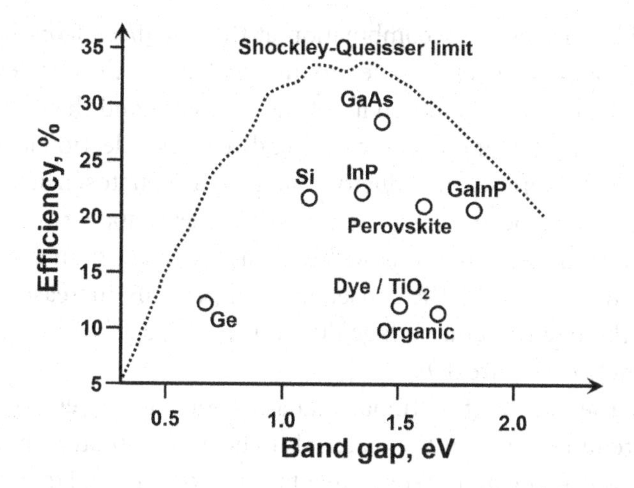

FIGURE 8.8 The Shockley-Queisser efficiency limit for single-junction solar cells and the efficiency of some available solar cells and solar cell materials. (The data are from [10] except for Ge.)

The Shockley-Queisser efficiency limit gives a quantitative rationale to the maximum theoretical efficiency of solar cells as a function of the band gap of the semiconductors used to construct them. Figure 8.8 shows such a dependence together with some examples of materials used in state-of-the-art PV devices. As one can see from the figure, the optimum band gaps are located close to approximately 1.4 eV. Notably, all the material and device efficiency examples are well below the Shockley-Queisser efficiency limits.

This nonideal behavior of the materials and devices originates from the losses in solar cells. The latter can be classified into the ones related to the following:

i. Recombination losses
ii. Metal/semiconductor contact losses
iii. Series resistance losses
iv. Reflection losses
v. Thermal losses

Recombination losses exist due to electron-hole recombination on the surface and in bulk, the recombination in the p-n junction, and various defects.

One should consider the recombination at the metal/semiconductor contact as well. Series resistance losses are unavoidable in electric devices and mainly depend on the conductivity of the materials and the design of the solar cells. A significant reflectance typically causes reflection losses in the spectral range where semiconductor materials are photosensitive. Finally, the thermal losses occur when the excessive energy appearing due to the absorption of the solar photons with the energy larger than semiconductor E_g is realized as heat. The consequent temperature increase effectively decreases the open-circuit voltage due to the “leakage” of carriers across the p-n junction (Figure 8.9).

Like in the case of functional materials for various energy applications, the requirements for the materials to be used in solar cells are rather straightforward. The main requirements for an optimal solar cell material are as follows:

a. A band gap between 1.1 and 1.7 eV

b. Consisting of readily available and nontoxic materials

c. Good PV conversion efficiency

d. Reasonable long-term stability

Silicon has many of the properties mentioned above. It is abundant, has a band gap of *ca* 1.1 eV, provides relatively long stability so that the commercial cells can be in operation during *ca* 20 years, and has an acceptable

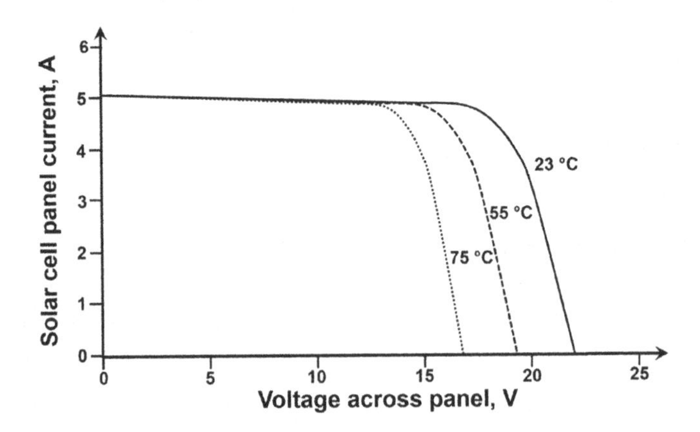

FIGURE 8.9 Typical change in the I-V characteristics of solar cell panels upon the increase of the cell temperature.

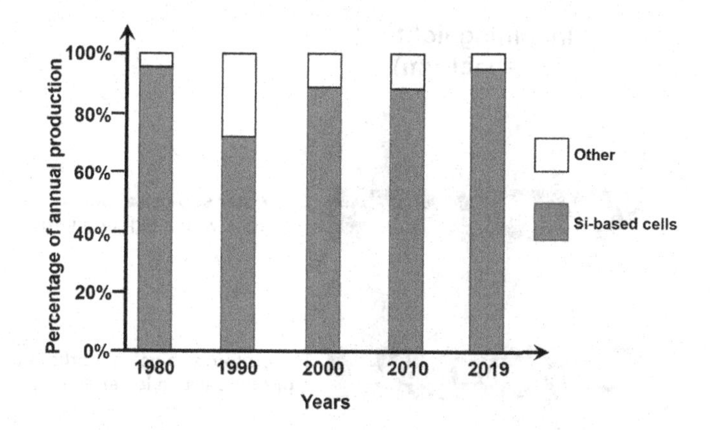

FIGURE 8.10 Solar cell production by technology. (The data are from [5].)

efficiency of 20%–25% [11], which can still be substantially improved (see Figure 8.8). Therefore, the most currently used commercial solar cells are Si-based solar cells (see Figure 8.10).

Among the Si-based cells, those produced using monocrystalline material provide higher efficiencies, primarily thanks to the lower content of defects in the crystal structure, which reduces the probability of electron-hole recombination. However, monocrystalline Si-based solar cells require an energy-demanding manufacturing process at high temperatures in the range between 400°C and 1400°C, as well as numerous lithographic processes. Therefore, despite a growing interest in Si-based commercial solar cells, a number of other approaches to increase the cell efficiency and/or reduce the overall costs of the units are in focus in the research and development worldwide, not only for large-scale common applications but also for space programs [12]. For instance, one approach to increase cell efficiencies is to use the so-called tandem cells, which are the subject of the next section.

8.3 TANDEM SOLAR CELLS

The multiple-junction, also known as tandem, solar cells have been developed to acquire higher efficiencies than those possible using just single p-n junction devices [13]. They were elaborated in the 1970s–1980s when the double junction cells were constructed using an AlGaAs-based architecture grown on top of a GaAs junction. The basic idea of these types of tandem cells is schematically explained in Figure 8.11. The outer subcell is built using a semiconductor with a wider band gap, E_{g1}. Such an outer cell aims to harvest photons with higher energy, $h\nu > E_{g1}$, while the

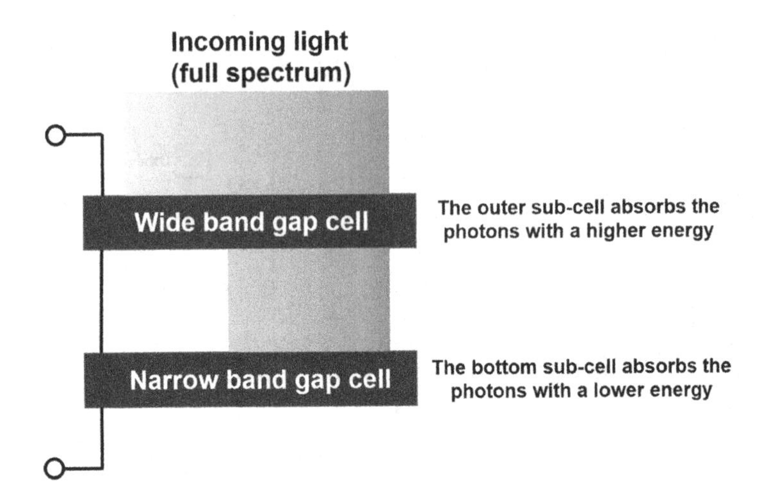

FIGURE 8.11 Basic principles of a tandem cell consisting of two p-n junctions and two semiconductors with different band gaps.

lower-energy photons, $hv < E_{g1}$, can go through without absorption. They reach the second cell made of a semiconductor with a narrower band gap, $E_{g2} < E_{g1}$, and the remaining photons having energy higher than E_{g2} are also absorbed, increasing the resulting efficiency of the PV device. In principle, one can continue adding more and more subcells with constantly decreasing band gap to make use of as many incoming photons as possible.

In Figure 8.11, only one concept of the cell connection is shown. Some other types of connections of individual subcells are illustrated in Figure 8.12 using three subcells as an example. They include independent, series, and hybrid connections. While independent and hybrid connections provide flexibility, approaches other than series-connected nowadays introduce additional cost and complexity into power electronics and solar cell design using existing technology. Connection in series is unfortunately also not ideal as it places substantial additional constraints on the tandem efficiency by limiting the materials, which can be used. Nevertheless, the series connection is currently the most popular approach.

Let us consider what a theoretical advantage in the efficiency of a two-junction solar cell is if one uses the well-accepted technology (Figure 8.13).

For the case of unconcentrated sunlight, a maximum efficiency of *ca* 42.3% was estimated for $E_{g1} \sim 1.9$ eV and $E_{g2} \sim 1.0$ eV [14]. Interestingly, the highest theoretical efficiency assessed to be approximately 86.8% if a multiple-junction cell would be realized, consisting of an infinite number of subcells, with smoothly changing band gaps and illuminated by

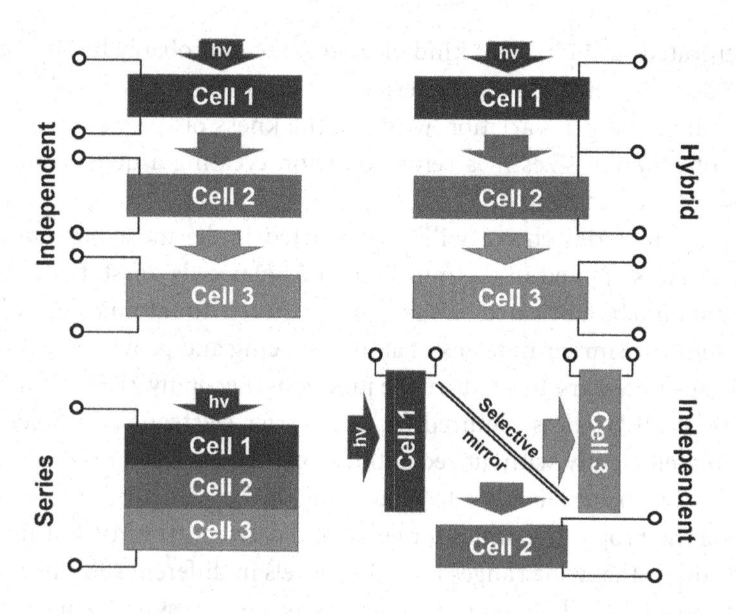

FIGURE 8.12 Different connections of individual cells in multiple junction solar cells: independent, series, and hybrid, as indicated in the Figure. Cell 1 should have the widest band gap, Cell 2 – intermediate band gap, and Cell 3 – the smallest band gap.

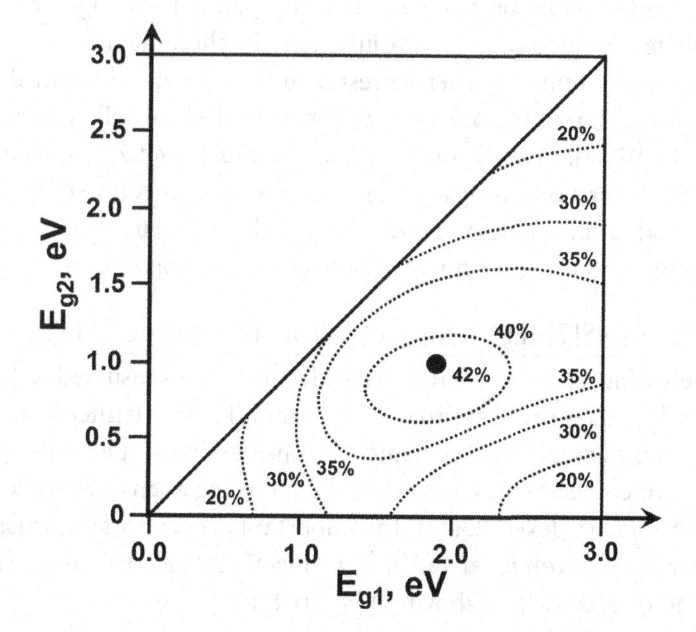

FIGURE 8.13 Efficiency limits as a function of the two semiconductor band gaps E_{g1} and E_{g2} for a two-p-n-junction tandem solar cell. (Adapted from [14].)

concentrated sunlight. This kind of solar cell can probably be approached if one uses a semiconductor material with a porosity gradient. The idea is to use the band gap variation with the thickness of the walls separating pores of different sizes in a semiconductor, creating a quasi-continuous change in the band gap [15].

Typical material choice will be restricted if the most popular series connection is applied in the multijunction solar cells. First, the complete device should be fabricated on the same substrate. The atomic device structures must be similar in terms of atomic spacing and provide the multiple band gaps necessary to produce the junctions. Secondly, the high material quality of all layers is required for the carrier electrons and holes to be collected efficiently without recombination. The next requirement is that high-quality materials need to be used as the so-called inactive layers to provide the proper dynamics for generated carriers. The last-but-not-least constraint is that wide ranges in doping levels in different subcells must be well controllable. This requirement exists as it is necessary to interconnect each subjunction in the stack, which requires reversing the n-p polarity between the subcells.

The III–V multijunction solar cells (III and V correspond to the groups in the periodic table of elements) became popular during the last few decades [16]. Semiconductor p-n junctions in these devices are epitaxially grown one upon another to result in 3 or 4 subcells with different band gaps. Around 40% of efficiency for such kinds of cells was achieved in *ca* 2012 [17,18]. Shortly afterward, an impressive 44.7% efficiency was reported [19] for a cell using a four-p-n-junction architecture (Figure 8.14), and 44%–46% efficiencies were demonstrated recently using four-junction and six-junction designs under sunlight concentration [20,21].

8.4 DYE-SENSITIZED AND PEROVSKITE SOLAR CELLS

An interesting class of solar cells called the dye-sensitized solar cells (DSSCs) [22] uses some approaches common for the artificial photosynthesis schemes described in Chapter 7. Consider an n-type semiconductor like TiO_2 in contact with an electrolyte containing a sensitizer redox complex, S^+/S, or complexes [23,24]. The important property of the sensitizer is that after the electron is excited, it can "inject" this electron to the conduction band of TiO_2 [25], as shown in Figure 8.15.

Instead of another type of semiconductor, a complex metal/transparent electron conductor counter electrode [26] is used to provide this electron to another redox pair I_3^-/I^- through an external electron conductor.

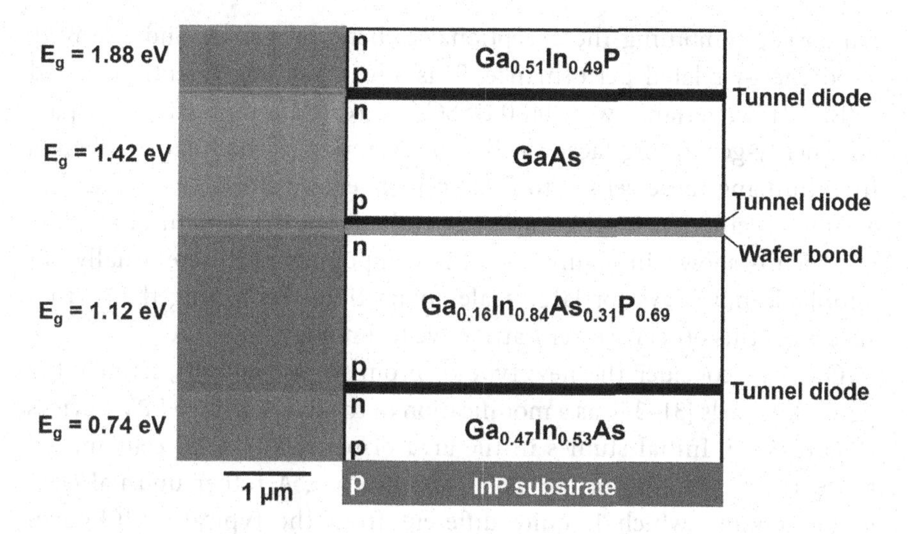

FIGURE 8.14 The schematic layer structure of the four-junction wafer-bonded solar cell (right) combined with the cross-section scanning electron microscopy image of the cell structure (left). (Adapted from [19].)

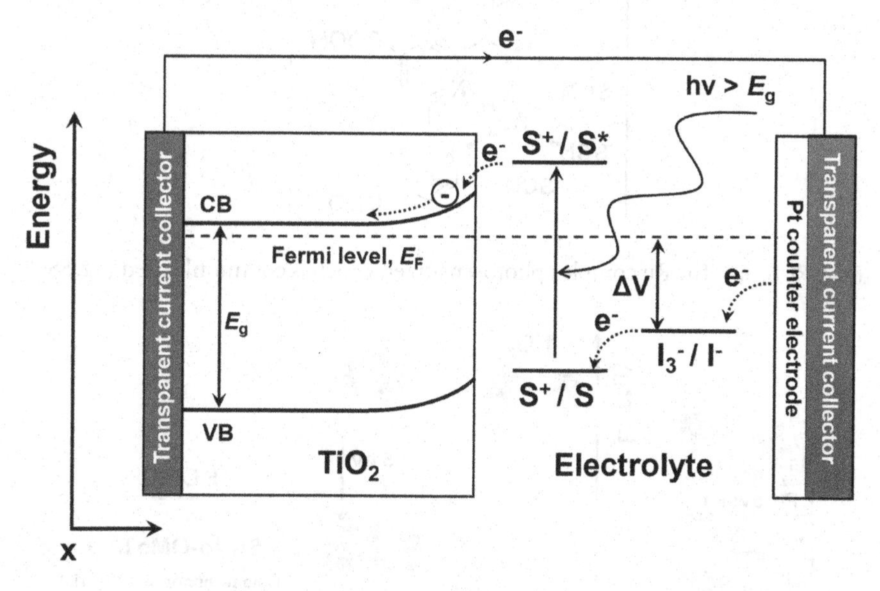

FIGURE 8.15 Schematic structure of a typical DSSC. S denotes a sensitizer electroactive complex.

The I_3^-/I^- redox pair spontaneously transfers it back to the S^+/S redox complex, as schematically shown in Figure 8.15, thus completing the cycle.

The efficiency of the DSSCs is *ca* 10%–12%. Probably, one of their main advantages is that they can be made transparent and used as "intelligent windows", combining the exceptional esthetic properties and relatively good energy-related performance. This advantage was recently realized in EPFL (Switzerland) with 1400 DSSCs to create an impressive transparent energy-generating façade [27]. However, one of the main challenges in optimizing these cells is to find an inexpensive alternative [28–30] to a quite specific sensitizer complex of ruthenium, the structural formula of which is shown in Figure 8.16. The compounds of Ru are usually not affordable nowadays for large-scale applications, restricting the further success of this otherwise-very-attractive technology.

One can consider the new type of promising solar cells, namely the perovskite cells [31–33], as a modification of the dye-sensitized PV devices (Figure 8.17). Initial studies in the area of perovskite solar cells indeed arose as an evolution of the DSSC-architecture. A rather unusual class of perovskites, which is quite different from the typical, well-known

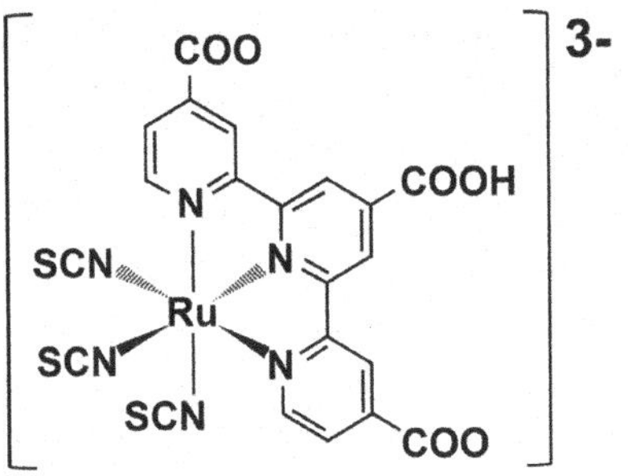

FIGURE 8.16 The Ru complex photosensitizer, which is commonly used in DSSCs.

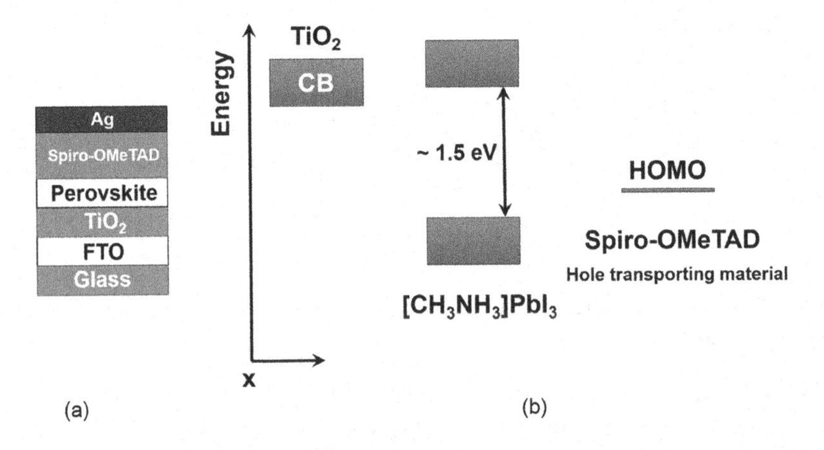

FIGURE 8.17 A schematic structure (a) of a perovskite solar cell with $[CH_3NH_3]PbI_3$ perovskite as a photoactive material with a respective energy diagram (b).

perovskite oxides, namely methylammonium tin and lead halides, attracted particular consideration in this respect [34]. The band gap for these compounds varies between ~1.57 eV and ~2.3 eV, depending on the halide content. Perovskite absorber materials with methyl-ammonium lead halides appeared nevertheless to be relatively cheap to manufacture. The crystal structure of the most used perovskite absorber $CH_3NH_3PbI_3$ (band gap ~1.55 eV) is shown in Figure 8.18.

Perovskite solar cells initiated in 2009 [35] with efficiencies of ~3.8% very quickly reached those over 20% within less than a decade afterward. Despite the challenges in the commercialization [36,37], the first products appeared in the market in 2021 [38]. The perovskite solar cells have achieved (2021) a lab-scale power conversion efficiency of ~25% [39], which is close to the performance of the best commercial silicon solar cells. With

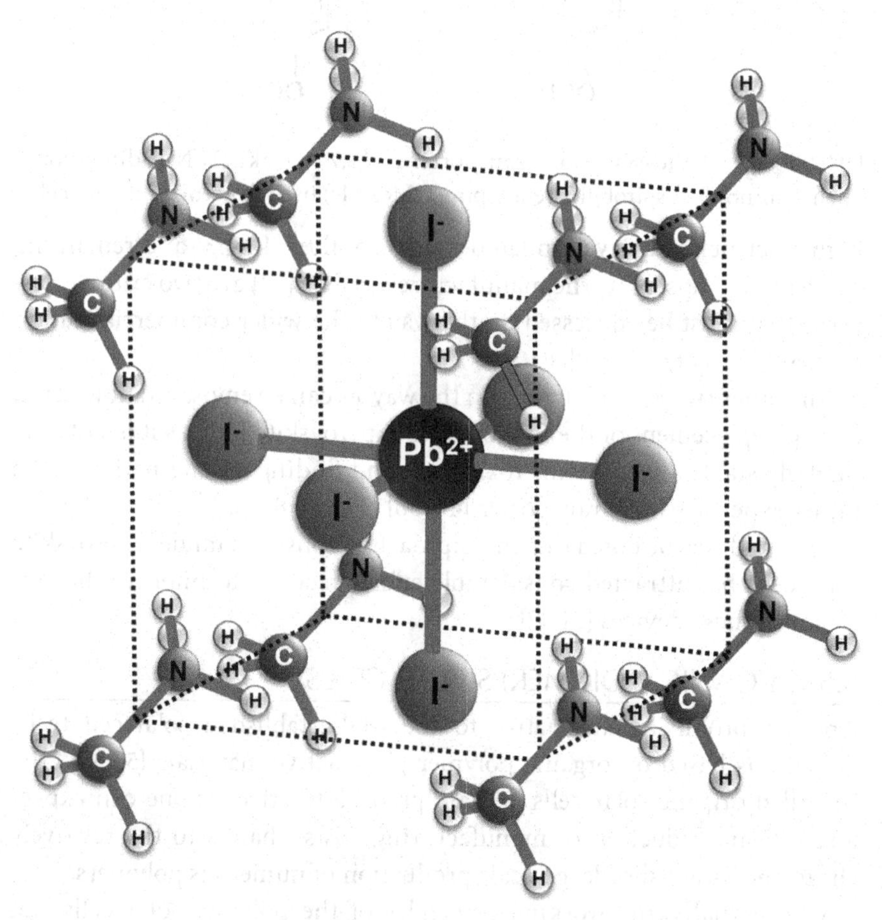

FIGURE 8.18 The crystal structure of the perovskite absorber $CH_3NH_3PbI_3$.

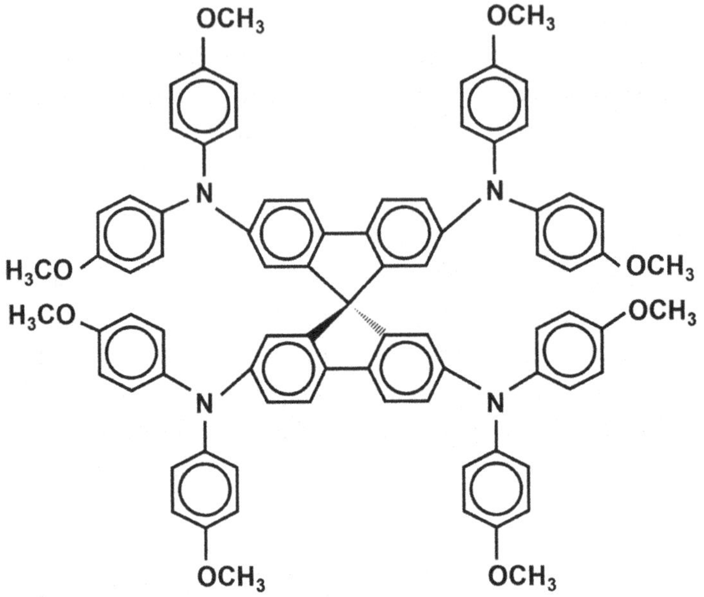

FIGURE 8.19 Solid-state hole transporter 2,2′,7,7′-tetrakis(N,N-p-dimethoxy-phenylamino)-9,9′-spirobifluorene (spiro-OMeTAD) used in perovskite solar cells.

high efficiencies achieved in lab devices, stability [40–43] and remaining challenges in upscaling the manufacture of PSCs [44] are two critical concerns that must be addressed on the way to the wider commercialization of these types of cells [45].

The other two major barriers on the way, as can be envisaged nowadays, are the replacement of the Pb-containing perovskite [46], as it is not particularly stable and contains toxic lead, and finding an alternative to the rather expensive hole transporter [47–50] (Figure 8.19).

It should also be noted that the approach to construct tandem perovskite solar cells has attracted considerable efforts recently to improve the efficiency of these devices [51,52].

8.5 ORGANIC (POLYMER) SOLAR CELLS

Another promising alternative to the well-established solar-cell technologies is based on organic polymer photoactive materials [53,54]. The so-called organic solar cells [55] are pretty attractive, as one can expect a significant reduction of manufacturing costs, thanks to the relatively cheap and established large-scale production of numerous polymers.

Conceptually, the working principles of the polymer solar cells can be described using similar approaches as in the case of semiconductor

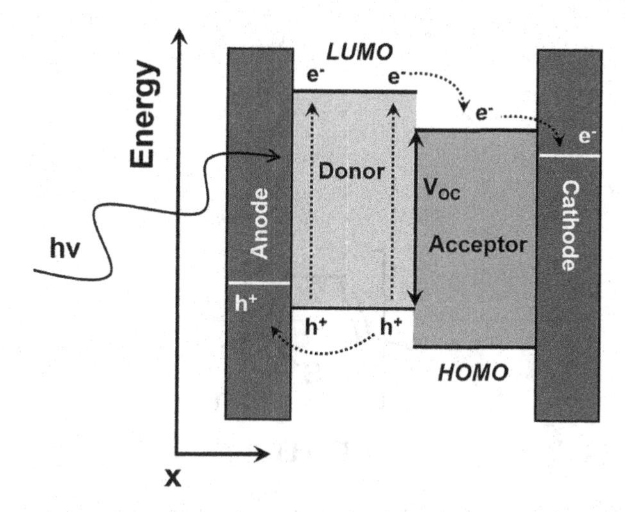

FIGURE 8.20 A schematic structure of a typical organic solar cell. HOMO – the highest occupied molecular orbital and LUMO – the lowest unoccupied molecular orbital.

materials (Figure 8.20). The main difference is that instead of the band gaps, one needs to use the concepts of the highest occupied molecular orbitals (HOMO) and lowest unoccupied molecular orbitals (LUMO). This is the consequence in the description of polymer systems, where the concept of molecular orbitals is often more adequate (Figure 8.20).

First, light enters the cell through the transparent anode, and it is absorbed in the bulk heterojunction layer through the generation of excitons. One should note that a molecule absorbs light when an electron is excited from the HOMO to the LUMO, as shown in Figure 8.20. The exciton has a relatively large binding energy (0.3–1 eV). Therefore, to produce photocurrent, the exciton must overcome it and dissociate into free electrons and holes. The generated excitons either recombine or reach a donor-acceptor interface, where they separate into electrons and holes. The electrons and holes move toward the anode and cathode through the donor and acceptor phases. So, one can see that similar concepts are used to construct the solar cell, just using different photoactive materials (Figure 8.20).

The efficiency of exciton diffusion depends on the exciton diffusion length (L_D), which should be short to avoid recombination.

$$L_D = (D\tau)^{0.5}$$

where D is the diffusion coefficient, and τ is the exciton lifetime.

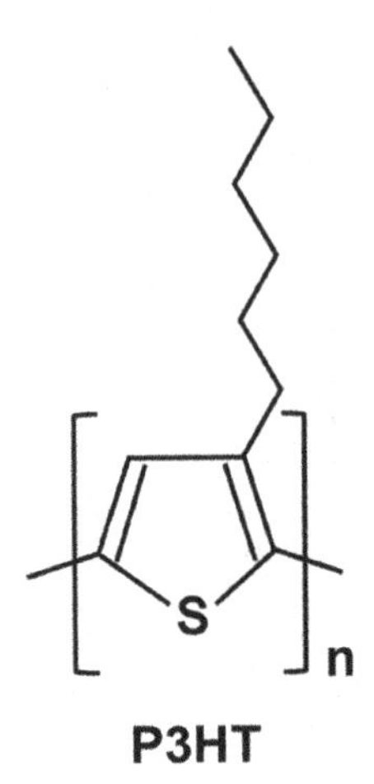

FIGURE 8.21 The structural formula of poly(3-hexylthiophene) (P3HT), the widely used polymer donor material for organic solar cells.

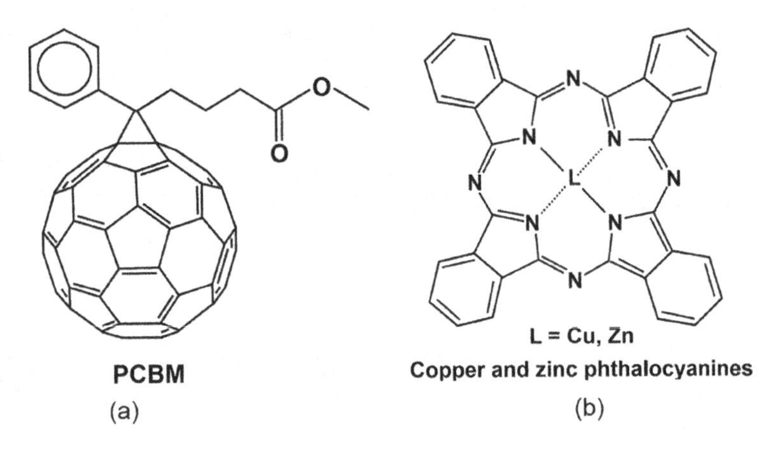

FIGURE 8.22 The structural formulas of (a) phenyl-C_{61}-butyric acid methyl ester, PCBM, and (b) metal phthalocyanines, common small-molecule acceptors for organic solar cells.

Typically, the exciton diffusion length in organic semiconductors ranges from 1 to 10 nm, limiting the absorbing layer thickness.

Challenges in advancing polymer solar cells include stability issues [56] and a relatively narrow choice of materials, especially acceptor materials [57–59]. The typical donor material for organic solar cells is poly(3-hexylthiophene) (P3HT, see Figure 8.21), while common small molecule acceptors are phenyl-C61-butyric acid methyl ester, PCBM, and metal phthalocyanines (Figure 8.22).

8.6 SUMMARY AND CONCLUSIONS

Efficient and affordable PV systems are extremely "expectable" in the current research and development related to the so-called renewable energy. From the current understanding, there is unprecedented diversity of ways of how to use this approach and contribute to the existing current schemes of energy provision or to design a variety of materials for photo-, and even electrochemical water splitting.

State-of-the-art materials for commercial solar cells include silicon with efficiencies exceeding 20% and a record of *ca* 25% for monocrystalline Si-cells in 2020. Depending on specific needs and applications, one can distinguish crystalline, amorphous, and porous Si-based materials for these devices. There are also families of materials for organic and other types of solar cells, including polymers, perovskites, TiO_2, GaAs, and other inorganic chalcogenide compounds. Approximately 7% efficiency for the transparent polymer solar and 12% efficiency for the nontransparent cells were reported. For the perovskite solar cells, the efficiencies in the range of 20%–25% seem to be easily achievable. However, some stability issues related to the organic and perovskite solar cells still prevent them from wider commercialization.

One of the most desirable aspects is to be able to easily tune the band gap and achieve good stability and flexibility in manufacturing. To partially address the first challenge, the tandem or multijunction solar cells are under development. The main difficulty here is selecting photoactive materials with appropriate band gaps, their position, and that the constituent materials are compatible with each other. Nevertheless, this field develops quickly with a clear promise to release a number of successful solutions for renewable energy provision systems.

8.7 QUESTIONS FOR SELF-CONTROL

1. What is the PV effect?
2. What are intrinsic semiconductors, and what are their schematic energy diagrams?
3. What are doped semiconductors, and what are the schematic energy diagrams in the case of p-doped and n-doped semiconductors?
4. Explain the space charge layer formation mechanisms at the semiconductor/electrolyte and semiconductor/semiconductor interfaces.
5. Explain what the p-n junction is.

6. Explain the working principles of Si-based PV cells.
7. What are the Shockley-Queisser efficiency limits?
8. What are the main requirements for ideal solar cell materials?
9. What are the common losses in solar cells, and how to minimize these losses?
10. What is the concept of multiple-junction (tandem) solar cells?
11. What are the requirements for the materials for PV applications considering tandem solar cells?
12. Tandem solar cells: what are the main difficulties in design and material choices?
13. What are the working principles of the polymer solar cells, and what are the key properties determining their performance?
14. Explain the working principles of dye-sensitized and perovskite solar cells.

REFERENCES

1. Becquerel, E. 1839. Memoire sur les effets electriques produits sous des rayons solaires. *Comptes rendus de l'Académie des Sciences* 9:561–567.
2. Einstein, A. 1905. Über einen die Erzeugung und erwandlung des Lichtes betreffenden heuristischen esichtspunkt. *Annalen der Physik* 17:132–148.
3. Ohl R.S. 1941. Light sensitive electric device. US Patent 2,402,622 (27 March).
4. Ohl R.S. 1941. Light sensitive electric device including silicon. US Patent 2,443,542 (27 May).
5. Photovoltaics Report. 2020. Fraunhofer Institute for Solar Energy Systems, Germany, www.ise.fraunhofer.de (Accessed: May 2021).
6. Solar PV Market Developments. United States Department of Energy, https://www.energy.gov/eere/solar/solar-energy-cost-analysis (Accessed: May 2021).
7. Badcock, J.; Lenzen, M. 2010. Subsidies for electricity-generating technologies: A review. *Energy Policy* 38:5038–5047.
8. Battersby, S. 2019. The solar cell of the future. *Proceedings of the National Academy of Sciences of the United States of America* 116:7–10.
9. Shockley, W.; Queisser, H.J. 1961. Detailed balance limit of efficiency of p-n junction solar cells. *Journal of Applied Physics* 32:510–519.
10. Polman, A.; Knight, M.; Garnett, E.C.; Ehrler, B.; Sinke W.C. 2016. Photovoltaic materials: Present efficiencies and future challenges. *Science* 352:aad4424.
11. Masuko, K.; Shigematsu, M.; Hashiguchi, T.; Fujishima, D.; Kai. M; Yoshimura, N.; Yamaguchi, T.; Ichihashi, Y.; Mishima, T.; Matsubara, N.;

Yamanishi, T.; Takahama, T.; Taguchi, M.; Maruyama, E.; Okamoto, S. 2014. Achievement of more than 25% conversion efficiency with crystalline silicon heterojunction solar cell. *IEEE Journal of Photovoltaics* 4:1433–1435.

12. Reb, L.K.; Böhmer, M.; Predeschly, B.; Grott, S.; Weindl, C.L.; Ivandekic, G.I.; Guo, R.; Dreißigacker, C.; Gernhäuser, R.; Meyer, A.; Müller-Buschbaum, P. 2020. Perovskite and organic solar cells on a rocket flight. *Joule* 4:1880–1892.
13. Li, X.; Xu, Q.; Yan, L.; Ren, C.; Shi, B.; Wang, P.; Mazumdar, S.; Hou, G.; Zhao, Y.; Zhang, X. 2021. Silicon heterojunction-based tandem solar cells: Past, status, and future prospects. *Nanophotonics* 10:2001–2022.
14. De Vos, A. 1980. Detailed balance limit of the efficiency of tandem solar cells. *Journal of Physics D: Applied Physics* 13:839–846.
15. Dzhafarov, T. 2013. Silicon solar cells with nanoporous silicon layer. Solar Cells - Research and Application Perspectives. IntechOpen. doi:10.5772/51593.
16. Cotal, H.; Fetzer, C.; Boisvert, J.; Kinsey, G.; King, R.; Hebert, P.; Yoona, H.; Karam, N. 2009. III–V multijunction solar cells for concentrating photovoltaics. *Energy & Environmental Science* 2:174–192.
17. Ermer, J.H.; Jones, R.K.; Hebert P.; Pien, P.; King, R.R.; Bhusari, D.; Brandt, R.; Al-Taher, O.; Fetzer, C.; Kinsey, G.S.; Karam, N. 2012. Status of $C3MJ^+$ and C4MJ production concentrator solar cells at Spectrolab. *IEEE Journal of Photovoltaics* 2:209–213.
18. King, R.R.; Bhusari, D.; Larrabee, D.; Liu, X.Q.; Rehder, E.; Edmondson, K.; Cotal, H.; Jones, R.K.; Ermer, J.H.; Fetzer, C.M.; Law, D.C.; Karam, N.H. 2012. Solar cell generations over 40% efficiency. *Progress in Photovoltaics: Research and Applications* 20:801–815.
19. Dimroth, F.; Grave, M.; Beutel, P.; Fiedeler, U.; Karcher, C.; Tibbits, T.N.D.; Oliva, E.; Siefer, G.; Schachtner, M.; Wekkeli, A.; Bett, A.W.; Krause, R.; Piccin, M.; Blanc, N.; Drazek, C.; Guiot, E.; Ghyselen, B.; Salvetat, T.; Tauzin, A.; Signamarcheix, T.; Dobrich, A.; Hannappel, T.; Schwarzburg, K. 2014. Wafer bonded four-junction GaInP/GaAs//GaInAsP/GaInAs concentrator solar cells with 44.7% efficiency. *Progress in Photovoltaics* 22:277–282.
20. France, R.M.; Geisz, J.F.; García, I.; Steiner, M.A.; McMahon, W.E.; Friedman, D.J.; Moriarty, T.E.; Osterwald, C.; Ward, J.S.; Duda, A.; Young, M.; Olavarria, W.J. 2016. Design flexibility of ultra-high efficiency 4-junction inverted metamorphic solar cells. *IEEE Journal of Photovoltaics* 6:578–583.
21. Geisz, J.F.; Steiner, M.A.; Jain, N.; Schulte, K.L.; France, R.M.; McMahon, W.E.; Perl, E.E.; Friedman, D.J. 2018. Building a six-junction inverted metamorphic concentrator solar cell. *IEEE Journal of Photovoltaics* 8:626–632.
22. O'Regan, B.; Grätzel, M. 1991. A low-cost, high-efficiency solar cell based on dye-sensitized colloidal TiO_2 films. *Nature* 353:737–40.
23. Cole, J.M.; Pepe, G.; Al Bahri, O.K.; Cooper, C.B. 2019. Cosensitization in dye-sensitized solar cells. *Chemical Reviews* 119:7279–7327.
24. Zeng, K.; Chen, Y.; Zhu, W.-H.; Tian, H.; Xie, Y. 2020. Efficient solar cells based on concerted companion dyes containing two complementary components: An alternative approach for cosensitization. *Journal of the American Chemical Society* 142:5154–5161.

25. Roose, B.; Pathak, S.; Steiner, U. 2015. Doping of TiO_2 for sensitized solar cells. *Chemical Society Reviews* 44:8326–8349.
26. Wu, J.; Lan, Z.; Lin, J.; Huang, M.; Huang, Y.; Fan, L.; Luo, G.; Lin, Y.; Xie, Y.; Wei. Y. 2017. Counter electrodes in dye-sensitized solar cells. *Chemical Society Reviews* 46:5975–6023.
27. EPFL's ampus has the world's first solar window. https://actu.epfl.ch/news/epfl-s-campus-has-the-world-s-first-solar-window/ (Accessed: June 2021).
28. Liang, M.; Chen, J. 2013. Arylamine organic dyes for dye-sensitized solar cells. *Chemical Society Reviews* 42:3453–3488.
29. Kumara, N.T.R.N.; Lim, A.; Lim, C.M.; Petra, M.I.; Ekanayake, P. 2017. Recent progress and utilization of natural pigments in dye sensitized solar cells: A review. *Renewable and Sustainable Energy Reviews* 78:301–317.
30. Calogero, G.; Bartolotta, A.; Di Marco, G.; Di Carlo, A.; Bonaccorso, F. 2015. Vegetable-based dye-sensitized solar cells. *Chemical Society Reviews* 44:3244–3294.
31. Kim, J.Y.; Lee, J.W.; Jung, H.S.; Shin, H.; Park, N.G. 2020. High-efficiency perovskite solar cells. *Chemical Reviews* 120:7867–7918.
32. Kausar, A.; Sattar, A.; Xu, C.; Zhang, S.; Kang, Z.; Zhang, Y. 2021. Advent of alkali metal doping: A roadmap for the evolution of perovskite solar cells. *Chemical Society Reviews* 50:2696–2736.
33. Jena, A.K.; Kulkarni, A.; Miyasaka, T. 2019. Halide perovskite photovoltaics: Background, status, and future prospects. *Chemical Reviews* 119:3036–3103.
34. Grätzel, M. 2014. The light and shade of perovskite solar cells. *Nature Materials* 13:838–842.
35. Kojima, A.; Teshima, K.; Shirai, Y.; Miyasaka, T. 2009. Organometal halide perovskites as visible-light sensitizers for photovoltaic cells. *Journal of the American Chemical Society* 131:6050–6051.
36. Rong, Y.; Hu, Y.; Mei, A.; Tan, H.; Saidaminov, M.I.; Seok, S.I.; McGehee, M.D.; Sargent, E.H.; Han, H. 2018. Challenges for commercializing perovskite solar cells. *Science* 361:eaat8235.
37. Petrus, M.L.; Schlipf, J.; Li, C.; Gujar, T.P.; Giesbrecht, N.; Müller-Buschbaum, P.; Thelakkat, M.; Bein, T.; Hüttner, S.; Docampo; P. 2017. Capturing the sun: A review of the challenges and perspectives of perovskite solar cells. *Advanced Energy Materials* 7:1700264.
38. Polish firm opens cutting-edge solar energy plant. https://www.france24.com/en/live-news/20210521-polish-firm-opens-cutting-edge-solar-energy-plant-1 (Accessed: June 2021).
39. Jeong, J.; Kim, M.; Seo, J.; Lu, H.; Ahlawat, P.; Mishra, A.; Yang, Y.; Hope, M.A.; Eickemeyer, F.T.; Kim, M.; Yoon, Y.J.; Choi, I.W.; Darwich, B.P.; Choi, S.J.; Jo, Y.; Lee, J.H.; Walker, B.; Zakeeruddin, S.M.; Emsley, L.; Rothlisberger, U.; Hagfeldt, A.; Kim, D.S.; Grätzel, M.; Kim, J.Y. 2021. Pseudo-halide anion engineering for α-$FAPbI_3$ perovskite solar cells. *Nature* 592:381–385.
40. Dong, C.; Xu, B.; Liu, D.; Moloney, E.G.; Tan, F.; Yue, G.; Liu, R.; Zhang, D.; Zhang, W.; Saidaminov, M.I. 2021. Carbon-based all-inorganic perovskite solar cells: Progress, challenges and strategies toward 20% efficiency. *Materials Today* 46. doi:10.1016/j.mattod.2021.05.016.

41. Vasilopoulou, M.; Fakharuddin, A.; Coutsolelos, A.G.; Falaras, P.; Argitis, P.; Yusoff, A.R.M.; Nazeeruddin, M.K. 2020. Molecular materials as interfacial layers and additives in perovskite solar cells. *Chemical Society Reviews* 49:4496–4526.
42. Boyd, C.C.; Cheacharoen, R.; Leijtens, T.; McGehee, M.D. 2019. Understanding degradation mechanisms and improving stability of perovskite photovoltaics. *Chemical Reviews* 119:3418–3451.
43. Zhou, C.; Tarasov, A.B.; Goodilin, E.A.; Chen, P.; Wang, H.; Chen, Q. 2022. Recent strategies to improve moisture stability in metal halide perovskites materials and devices. *Journal of Energy Chemistry* 65:219–235.
44. Schulz, P.; Cahen, D.; Kahn, A. 2019. Halide perovskites: Is it all about the interfaces? *Chemical Reviews* 119:3349–3417.
45. Li, N.; Niu, X.; Chen, Q.; Zhou, H. 2020. Towards commercialization: the operational stability of perovskite solar cells. *Chemical Society Reviews* 49:8235–8286.
46. Chen, M.; Kamarudin, M.A.; Baranwal, A.K.; Kapil, G.; Ripolles, T.S.; Nishimura, K.; Hirotani, D.; Sahamir, S.R.; Zhang, Z.; Ding, C.; Sanehira, Y.; Bisquert, J.; Shen, Q.; Hayase, S. 2021. High-efficiency lead-free wide band gap perovskite solar cells via guanidinium bromide incorporation. *ACS Applied Energy Materials* 4. doi:10.1021/acsaem.1c00413.
47. Urbani, M.; de la Torre, G.; Nazeeruddin, M.K.; Torres, T. 2019. Phthalocyanines and porphyrinoid analogues as hole- and electron-transporting materials for perovskite solar cells. *Chemical Society Reviews* 48:2738–2766.
48. Yin, X.; Song, Z.; Li, Z.; Tang, W. 2020. Toward ideal hole transport materials: a review on recent progress in dopant-free hole transport materials for fabricating efficient and stable perovskite solar cells. *Energy and Environmental Science* 13:4057–4086.
49. Heidariramsheh, M.; Mirhosseini, M.; Abdizadeh, K.; Mahdavi, S.M., Taghavinia, N. 2021. Evaluating Cu_2SnS_3 nanoparticle layers as hole-transporting materials in perovskite solar cells. *ACS Applied Energy Materials* 4:5560–5573.
50. Li, Y.; Wu, F.; Han, M.; Li, Z.; Zhu, L.; Li, Z. 2021. Merocyanine with hole-transporting ability and efficient defect passivation effect for perovskite solar cells. *ACS Energy Letters* 6:869–876.
51. Li, H.; Zhang, W. 2020. Perovskite tandem solar cells: From fundamentals to commercial deployment. *Chemical Reviews* 120:9835–9950.
52. Yan, C.; Huang, J.; Li, D.; Li, G. 2021. Recent progress of metal-halide perovskite-based tandem solar cells. *Materials Chemistry Frontiers* 5: 4538–4564.
53. Mazzioa, K.A.; Luscombe, C.K. 2015. The future of organic photovoltaics. *Chemical Society Reviews* 44:78–90.
54. Hedley, G.J.; Ruseckas, A.; Samuel, I.D.W. 2017. Light harvesting for organic photovoltaics. *Chemical Reviews* 117:796–837.
55. Marinova, N.; Valero, S.; Delgado J.L. 2017. Organic and perovskite solar cells: Working principles, materials and interfaces. *Journal of Colloid and Interface Science* 488:373–389.

56. Cheng, P.; Zhan, X. 2016. Stability of organic solar cells: Challenges and strategies. *Chemical Society Reviews* 45:2544–2582.
57. Wadsworth, A.; Moser, M.; Marks, A.; Little, M.S.; Gasparini, N.; Brabec, C.J.; Baran, D.; McCulloch, I. 2019. Critical review of the molecular design progress in non-fullerene electron acceptors towards commercially viable organic solar cells. *Chemical Society Reviews* 48:1596–1625.
58. Zhang, G.; Zhao, J.; Chow, P.C.Y.; Jiang, K.; Zhang, J.; Zhu, Z.; Zhang, J.; Huang, F.; Yan, H. 2018. Nonfullerene acceptor molecules for bulk heterojunction organic solar cells. *Chemical Reviews* 118:3447–3507.
59. Zhao, Z.W.; Omar, Ö.H.; Padula, D.; Geng, Y.; Troisi, A. 2021. Computational identification of novel families of nonfullerene acceptors by modification of known compounds. *The Journal of Physical Chemistry Letters* 12:5009–5015.

CHAPTER 9

Transparent Electron Conductors

9.1 MOTIVATION

DEFINITION:

When defining the transparent electron conductors, one usually takes into account that they should have high optical transmission at visible wavelengths and electron conductivity close to or comparable to that of metals.

Transparent electron conductors are widely used in energy provision devices. In particular, one can mention solar cells of different types like silicon solar cells or dye-sensitized solar cells as well as some "artificial leaves". In fact, transparent electron conductors are considered as extremely important key functional materials in these applications. They also found their niches in hybrid energy devices, which use sunlight as a source of energy. Notably, these materials are widely used in mobile devices to form, for example, touch screens.

One can distinguish at least three classes of such materials. These are (i) oxides, (ii) transparent conducting polymers, and (iii) polymer/metal composites. In the following, let us consider how a material can have both transparency and high electronic conductivity.

9.2 OXIDE MATERIALS

The first class of electronically conducting transparent materials comprises oxides [1–5]. The earliest work about a transparent electron oxide conductor, CdO, appeared in the literature in 1907 [6]. By now, several oxides with

DOI: 10.1201/9781003025498-9

similar properties have been identified [7], including the so-called binary oxides like In_2O_3/SnO_2 (indium-doped tin oxide (ITO)), SnO_2/F (fluorine-doped tin oxide (FTO)), ZnO, and ternary oxides (combination of binary oxides) such as Cd_2SnO_4, $CdSnO_3$, Zn_2SnO_4, $CdIn_2O_4$, $Zn_2In_2O_5$, $MgIn_2O_4$, or $In_4Sn_3O_{12}$. Recently, FTO has been particularly recognized because it is relatively stable under atmospheric conditions, chemically inert, temperature resistant, mechanically stable, and cheaper than, e.g., indium tin oxide. On the other hand, the transmittance in the visible region is higher in the case of the more expensive ITO films than for the FTO layers.

From the point of view of the band theory of solids, as discussed in previous chapters, transparent (for visible light) materials would require band gaps (E_g) between *ca* 1.6 eV and 3.3 eV. Materials typically demonstrate a relatively good electron conductivity if they have an E_g of *ca* 0.45 eV (e.g., tellurium) or less. From these first simple considerations, materials can be either transparent or good electron conductors. So, how can a material display both properties?

First of all, it is essential to start with a transparent oxide (n-type) material with a band gap of $E_g > 3$ eV and high optical transparency ($T \geq 85\%$). One then performs doping, i.e., replaces the O atoms in pure tin oxide, SnO_2, with F atoms to form FTO. When the amount of F atoms is *ca* 10%, the electron carrier concentration is very high. Basically, one creates the so-called "degenerate semiconductor", i.e., a material, which starts to act more like a metal than as a semiconductor due to the high doping level. As the doping concentration increases, electrons populate states within the conduction band, pushing the Fermi level higher in energy. In the case of a degenerate doping level, the Fermi level lies inside the conduction band (Figure 9.1).

In the case of a degenerate semiconductor, an electron from the top of the valence band can only be excited by incoming photons of appropriate energy into the conduction band above the Fermi level (which now lies in the conduction band) since all the states below the Fermi level are occupied states. Thus, one can also observe an increase in the apparent band gap, the so-called Burstein-Moss's shift (see Figure 9.1). Therefore, there are simultaneously two effects that exist in such a case, as described below.

- The Fermi level lies in the conduction band, and the electronic conductivity is high, even at room temperature
- The apparent band gap is wide enough that the material appears transparent in the visible region

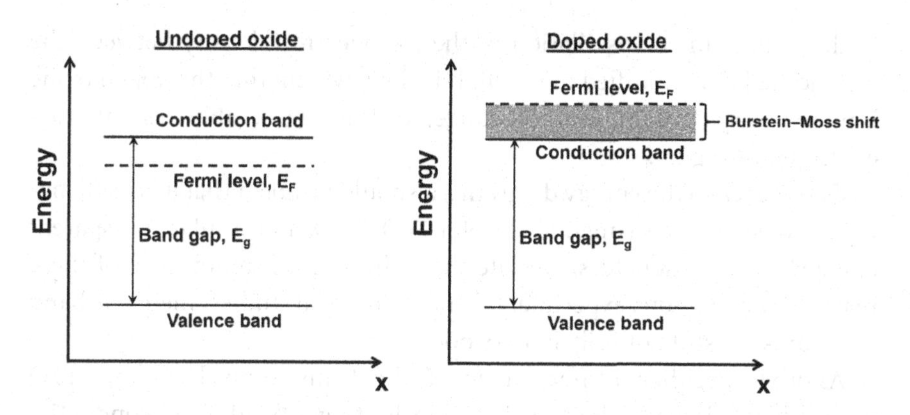

FIGURE 9.1 A schematic energy diagram for intrinsic undoped and doped (degenerate) n-type oxide semiconductors. Burstein-Moss's shift is also schematically explained.

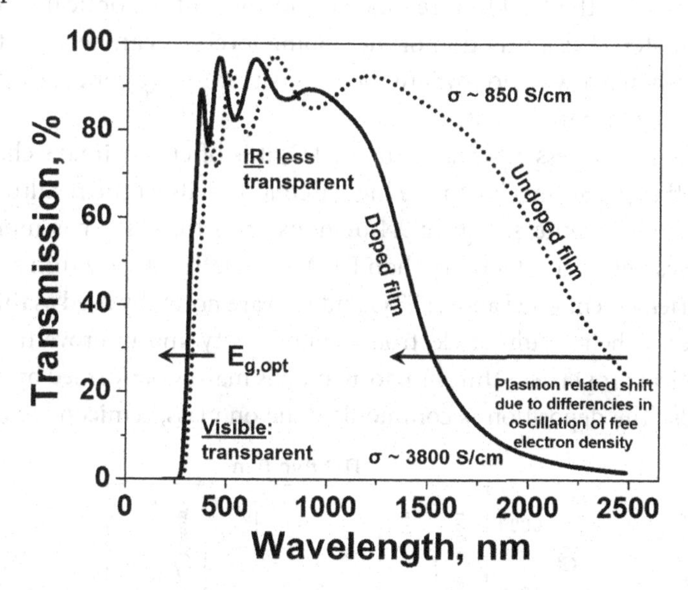

FIGURE 9.2 Realistic optical transmission spectra of undoped and ITO materials. (Adapted from [2].)

Relatively good stability makes some of the transparent electron conductors key functional materials for use in, e.g., solar cells. Figure 9.2 shows realistic transmission spectra of doped and undoped oxides with electron conductivities of *ca* 3800 S/cm and 850 S/cm, respectively [8]. The Burstein-Moss's shift manifests itself in the spectra at a lower wavelength, as indicated in the Figure as $E_{g,\,opt}$. At a higher wavelength, one can observe a drastic shift after the doping. This is a plasmon-related shift resulting in differences in the oscillation of the free electron density between the unmodified and modified material. It is then evident that the doped oxide is more transparent in the visible range, while it is noticeably less transparent in the IR region.

Going back to the energy diagrams, it should be noted that it is challenging to construct experimental ones for ITO, FTO, and similar transparent oxide electron conductors. Despite the technological importance of these materials and various experimental and theoretical studies, their fine band structures are still not fully understood.

Another peculiar characteristic of the transparent FTO and ITO materials is related to the fact that in order to get the desired conductivity and transparency, the doping level is mainly determined empirically. For instance, Figure 9.3 schematically shows the experimentally revealed dependence of the ITO film resistance and some of the optical properties of this material as a function of the doping level. As can be seen, there is a relatively narrow region where the film is maximally conducting and at the same time transparent.

While ITO is less affordable than FTO, it is often a primary choice for solar cell applications due to the higher conductivity of the resulting films of the same thickness. Figure 9.4 demonstrates that ITO has more than two times better conductivity than FTO, especially for very thin samples.

As briefly mentioned above, FTO and ITO are normally used as thin films to increase their resulting electronic conductivity and improve the necessary optical properties. Thin-film formation is mainly performed by physical methods. The deposition is commonly done on glass, semiconductors, and

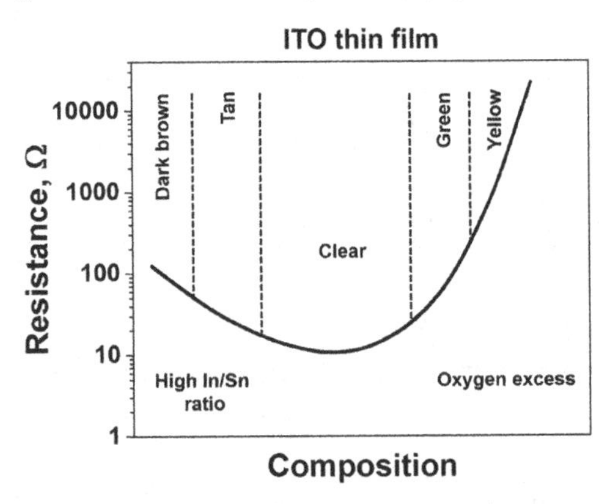

FIGURE 9.3 Schematic dependence of resistive and some optical properties of an ITO film as a function of composition.

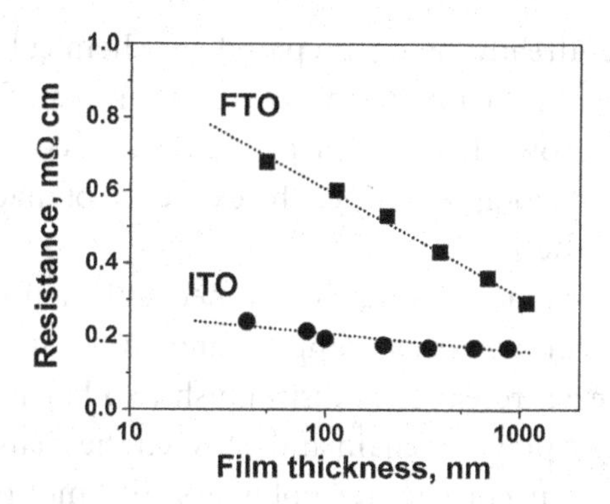

FIGURE 9.4 Dependence of ITO and FTO film resistance as a function of the respective film thickness. (The data are from [14].)

dielectric polymers. One can distinguish the following standard physical and physicochemical methods to do so [9,10]:

i. Magnetron sputtering techniques [11]
ii. Ion beam sputtering
iii. Reactive thermal and electron beam evaporation, etc.
iv. Spray pyrolysis [12]
v. Sol-gel methods [13]

While the ITO and FTO transparent electronic conductors are stable and have relatively good conductivities, they are still rather expensive due to issues such as, e.g., availability of indium or the challenges in thin film formation approaches. With the total global market for these two oxides approaching approximately two billion US dollars nowadays, they are probably not sustainable for thin-film photovoltaic applications at a significantly larger scale if compared with the current energy demand. Therefore, there is a growing interest in finding more affordable and cheap alternatives, for instance, using polymers or hybrid composite materials, as discussed in the following sections.

9.3 TRANSPARENT CONDUCTING POLYMERS

The 2000 Nobel Prize in Chemistry was awarded to Heeger, MacDiarmid, and Shirakawa for *the discovery and development of electrically conductive polymers*. However, important representatives of these materials, e.g., polyaniline, were already being prepared by chemical or even electrochemical oxidation in the nineteenth century. One can find that conducting polymers are known from at least 1862 [15]. Of course, they were not called polymers for a long time since the existence of macromolecules was not accepted until the 1920s.

It should be noted that the conductivity mechanism in electronically conducting polymers can be rather complex, and so far, no single model can represent it. However, researchers distinguish two key processes: the electron transfer along a polymer chain and between the chains (Figure 9.5).

In other words, in conducting polymers, the motion of delocalized electrons often occurs through conjugated systems. However, the electron hopping mechanism is also likely to be operative, especially between chains (interchain conduction) and defects. Interestingly, pure polymers with defect-free chains consisting of structural units with conjugated bonds are very often dielectric. Special doping is normally required to increase their conductivity. One can, in general, regulate the electrical conductivity and even the color of the transparent polymer conductors by controlling the oxidation state of the constituting atoms, polymer doping level, and morphology [16].

One typical example is polyaniline. Different states of this polymer correspond to its different conductivities and colors [17] as schematically shown in Figure 9.6. Unmodified polyaniline appears clear and colorless, with

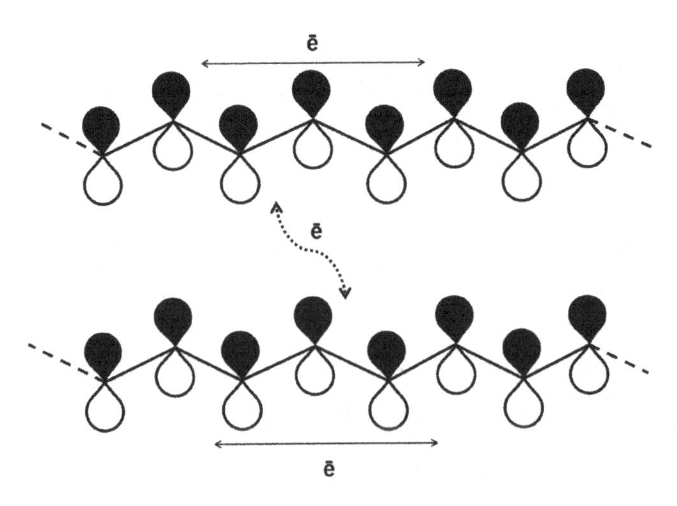

FIGURE 9.5 Schematic simplified model of electron transfer in electronically conducting polymers. One can consider electron transfer along a polymer chain with, e.g., conjugated bonds and between the chains.

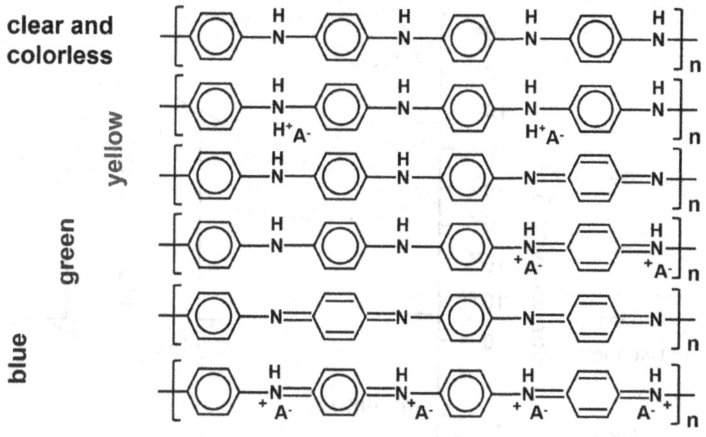

FIGURE 9.6 Structural formulas of polyaniline and its derivatives with different conductivities and optical properties. H^+ and A^- designate protons and anions, respectively.

moderate conductivity. However, by oxidizing certain carbon atoms in the chain or/and by protonating the nitrogen atoms, one can change not only the conductivity but also the color, for example, to yellow, green, or blue [18].

It is also remarkable that by changing the doping level of polyaniline, one can vary its electron conductivity in a very wide range, from those values typical for dielectrics (e.g., diamond) to semiconductors (e.g., Si or Ge) and metals (e.g., In or Sn) (see Figure 9.7).

Electronically conducting polymers can be successfully used in various energy provision devices and especially in photovoltaic applications [19,20]. For instance, uniform and transparent polyaniline electrodes can be used as counter electrodes in solar cells, particularly in the bifacial dye-sensitized solar cells [21] together with FTO or ITO electrodes, as schematically shown in Figure 9.8. In the latter case, since sunlight can penetrate the device from the front and the rear sides simultaneously, more dye species can be excited. This results in a higher number of generated charge carriers and, therefore, in the improvement of the overall efficiency and enhanced photovoltaic performance of these types of solar cells [22]. They also have good aesthetic properties as they are not only transparent but also often colorful and can be used in urban areas for applications in windows and building-integrated photovoltaics [23].

Despite a number of unique properties of electronically conductive transparent polymers, predicting their exact optical and electrical properties is still complicated. Therefore, more straightforward approaches based on

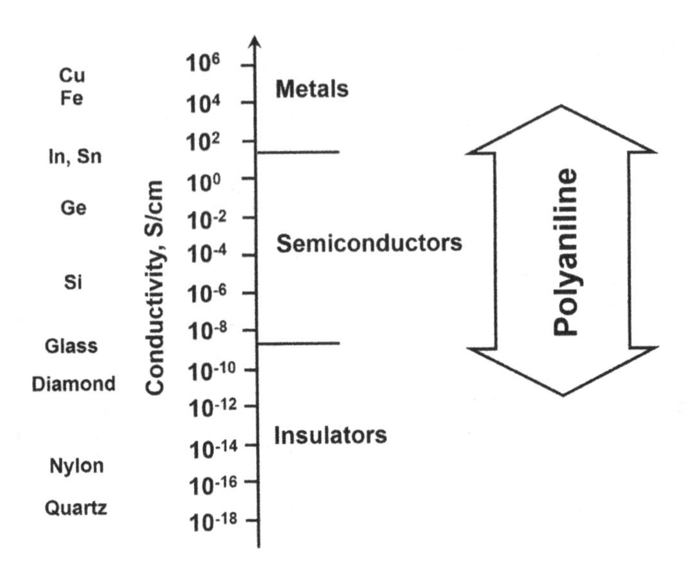

FIGURE 9.7 Depending on the doping level of polyaniline, its electronic conductivity can change from the one typical for insulators to the values common for metallic Sn or Fe.

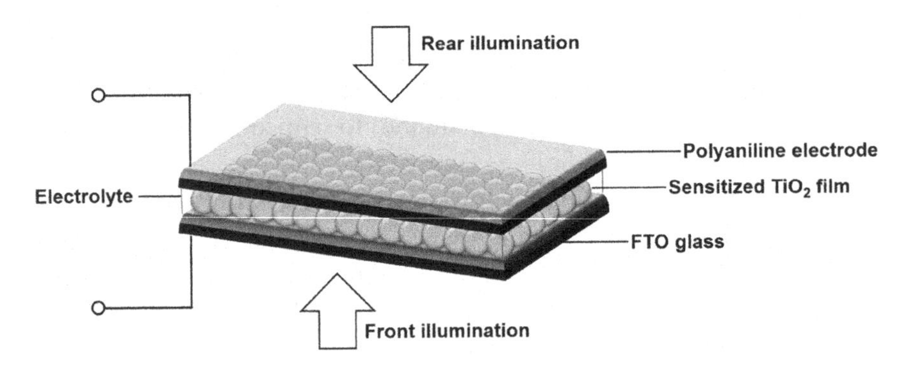

FIGURE 9.8 A sketch of a bifacial dye-sensitized solar cell with electronically conducting transparent FTO and polyaniline electrodes. (Adapted from [24].)

some composites and hybrid materials have recently been under broad investigation. In the following section, approaches using transparent structures consisting of polymer matrixes and metal particles are briefly considered.

9.4 HYBRID AND COMPOSITE MATERIALS

Design and implementation of transparent composite or hybrid conductors are probably one of the simplest approaches to predict the resulting optical properties, chemical, photo, and mechanical stability, and the

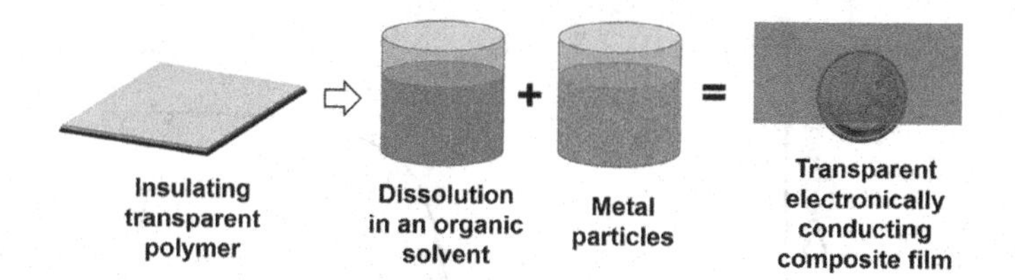

FIGURE 9.9 A general approach to prepare electronically conducting transparent hybrid films consisting of polymer matrix/metal particles.

electrical resistance of the materials [25]. A schematic describing one of the most straightforward procedures to prepare, for instance, thin composite films is shown in Figure 9.9.

Many insulating transparent conductors with improved mechanical properties and stability against degradation under atmospheric conditions and illumination are soluble in a number of organic solvents. This fact enables the preparation of polymer/solvent liquid phases to be mixed with suspensions containing the same solvent and small metal (e.g., Cu or Au) or even carbon nanotube [26] particles. After such mixing, the solvent can be evaporated, and a composite consisting of the solid transparent polymer matrix and incorporated metal particles can be obtained. If the amount of particles is high enough to enable the electronic conductivity and at the same time small enough to maintain good transparency, the material demonstrates similar properties to those for the oxides or polymers described in the previous sections. In order to find the optimum content of the particles balancing these two basic properties, the so-called percolation threshold should be determined and controlled.

DEFINITION:

In materials science, the term *percolation* is used when dealing with complex networks. Such networks can consist of, for instance, foreign particles in an isotropic or anisotropic media. One of the main outputs of the percolation theory for material scientists is the prediction of the so-called *percolation threshold.* At this threshold, the content of particles is close to that enabling the formation of long-range connectivity of, e.g., particles in random systems.

Figure 9.10 schematically shows the dependence of the measured electronic conductivity of a composite sample consisting of a polymer matrix and spherical metallic species. Initially, when the content of the particles is relatively low, they are normally not in contact with each other and do not form any steady networks. This results in almost no changes in the net

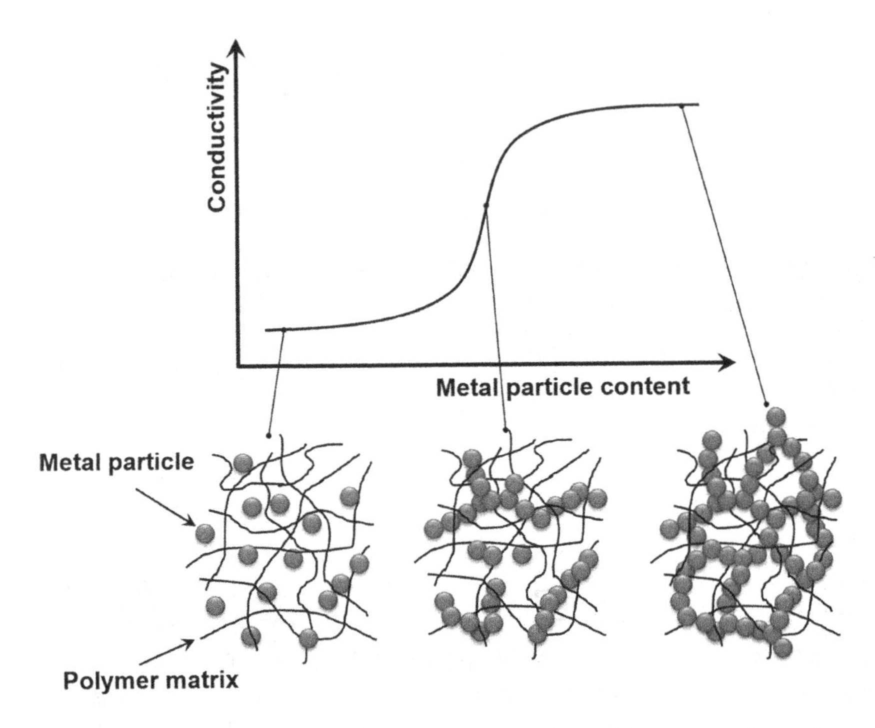

FIGURE 9.10 A schematic conductivity vs. metal particle content dependence of a sample consisting of a transparent insulating polymer (matrix) and spherical metallic species (filler). The very pronounced growth of the conductivity is observed at the percolation threshold.

conductivity. At the percolation threshold, the content of metal species is high enough to start developing the first metal "nonbreaking" chains, significantly enhancing the conductivity.

Finally, if one continues to increase the particle content, the newly formed metal chains will not increase the conductivity drastically, and a plateau in the curve shown in Figure 9.10 appears.

In order to keep the optical transparency in the visible region as high as possible, it is essential to shift the percolation threshold toward lower values. To do so, one can optimize the metal particle's size and shape. For example, metal nanowires [27,28] enable a high probability to form a continuous electronically conducting network at pretty low metal content providing high conductivities and transparency (see Figure 9.11) [29].

Alternative methods to form metal/polymer conducting and transparent electrodes include the so-called photonic sintering [30], which is high-temperature processing using, e.g., pulsed light sources and other similar approaches.

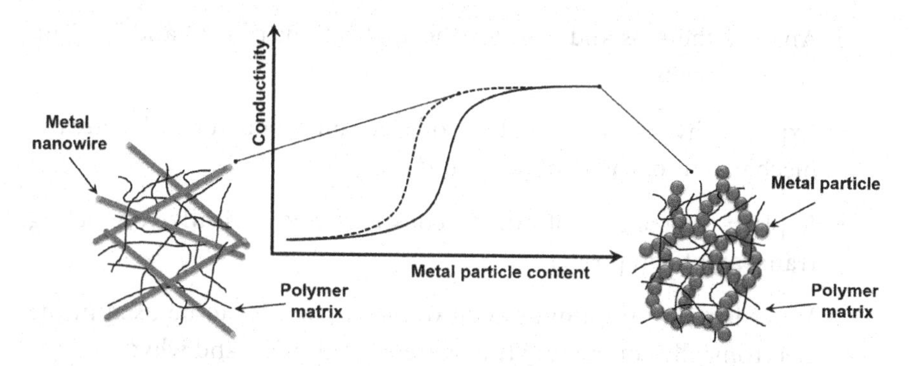

FIGURE 9.11 A schematic conductivity vs. metal particle content dependence of a sample consisting of a transparent insulating polymer and either spherical metallic species or metal nanowires (filler). The shape and also the size of the particles can control the position of the percolation threshold.

9.5 SUMMARY AND CONCLUSIONS

Transparent electron conductors represent a very important class of functional energy materials, and in many cases, they determine the successful scaling up of some energy provision technologies. The applications of transparent electron conductors range from solar cells or "artificial leaves" to various portable devices. The main classes of transparent electron conductors involve oxide materials, polymers, and composite or hybrid materials. As the state-of-the-art conductors, one can distinguish ITO, FTO, polyaniline, and systems, which involve insulating transparent polymer matrix with incorporated metal nanoparticles with their content above the percolation threshold, when the electron conductivity appeared to be high due to metal particles connected with each other while the transparency is still acceptable as the content of the metal is low enough. One can envisage that electronically conducting transparent polymers and composites will be more important as they are often more affordable than oxides for large-scale energy applications.

9.6 QUESTIONS

1. Name typical classes of transparent electron conductors.
2. Using energy diagrams, explain why oxide materials can have high electronic conductivity but remain transparent.
3. Name state-of-the-art transparent electronically conducting oxide materials.

4. Analyze the *pros* and *cons* for the applications of ITO and FTO in, e.g., solar cells.
5. Explain why polymers can be good electron conductors; what are the mechanisms of their conductivity?
6. Explain the concept of how to construct electronically conducting transparent composites.
7. What is the minimum amount of metal particles to be used in the electronically conducting transparent composites and why?

REFERENCES

1. Klein, A. 2013. Transparent conducting oxides: Electronic structure-property relationship from photoelectron spectroscopy with in situ sample preparation. *Journal of the American Ceramic Society* 96:331–345.
2. Stadler, A. 2012. Transparent conducting oxides. An up-to-date overview. *Materials* 5:661–683.
3. Song, J.Z.; Kulinich, S.A.; Li, J.H.; Liu, Y.L.; Zeng, H.B. 2015. A general one-pot strategy for the synthesis of high-performance transparent-conducting-oxide nanocrystal inks for all-solution-processed devices. *Angewandte Chemie International Edition* 54:462–466.
4. Jansons, A.W., Koskela, K.K.; Crockett, B.M.; Hutchison, J.E. 2017. Transition metal-doped metal oxide nanocrystals: Efficient substitutional doping through a continuous growth process. *Chemistry of Materials* 29:8167–8176.
5. Ginley, D.S., Hosono, H.; Paine, D.C. 2011. *Handbook of Transparent Conductors.* Springer: Berlin.
6. Bädeker, K. 1907. Über die elektrische Leitfähigkeit und die thermoelektrische Kraft einiger Schwermetallverbindungen. *Annalen der Physik* 327:749–766.
7. Hautier, G.; Miglio, A.; Ceder, G.; Rignanese, G.M.; Gonze, X. 2013. Identification and design principles of low hole effective mass p-type transparent conducting oxides. *Nature Communications* 4:2292.
8. Klein, A. 2012. Transparent conducting oxides, *Journal of American Ceramic Society* 96:1–15.
9. Duta, M.; Anastasescu, M.; Calderon-Moreno, J.M.; Predoana, L.; Preda, S.; Nicolescu, M.; Stroescu, H.; Bratan, V.; Dascalu, I.; Aperathitis, E.; Modreanu, M.; Zaharescu, M.; Gartner, M. 2016. Sol-gel versus sputtering indium tin oxide films as transparent conducting oxide materials. *Journal of Materials Science: Materials in Electronics* 27:4913–4922.
10. Chen, Z.X.; Li, W.C.; Li, R.; Zhang, Y.F.; Xu, G.Q.; Cheng, H.S. 2013. Fabrication of highly transparent and conductive indium-tin oxide thin films with a high figure of merit via solution processing. *Langmuir* 29: 13836–13842.

11. Kim, M.J.; Song, P.K. 2016. High crystallization of ultra-thin indium tin oxide films prepared by direct current magnetron sputtering with post-annealing. *Science of Advanced Materials* 8:622–626.
12. Manoj, P.K.; Joseph, B.; Vaidyana, V.K.; Amma, D.S.D. 2007. Preparation and characterization of indium-doped tin oxide thin films. *Ceramics International* 33:273–278.
13. Korosi, L.; Scarpellini, A.; Petrik, P.; Papp, S.; Dekany, I. 2014. Sol-gel synthesis of nanostructured indium tin oxide with controlled morphology and porosity. *Applied Surface Science* 320:725–731.
14. Minami, T. 2013. Chapter Five - transparent conductive oxides for transparent electrode applications. *Semiconductors and Semimetals* 88:159–200.
15. Letheby, M.B. 1862. On the production of a blue substance by the electrolysis of sulphate of aniline. *Journal of Chemical Society* 15:161–163.
16. Inzelt, G. 2008. *Conducting Polymers*. Springer. ISBN 978-3-540-75929-4.
17. Small, W.R.; Masdarolomoor, F.; Wallace, G.G.; in het Panhuis, M. 2007. Deposition and characterization of transparent conducting electroactive polyaniline composite films with a high carbon nanotube loading fraction. *Journal of Materials Chemistry* 17:4359–4361.
18. Mažeikienė, R.; Niaura, G.; Malinauskas, A. 2019. A comparative multi-wavelength Raman spectroelectrochemical study of polyaniline: A review. *Journal of Solid State Electrochemistry* 23:1631–1640.
19. Hou, W.; Xiao, Y.; Han, G.; Lin, J.Y. 2019. The applications of polymers in solar cells: a review. *Polymers* 11:143.
20. Ibanez, J.G.; Rincón, M.E.; Gutierrez-Granados, S.; Chahma, M.; Jaramillo-Quintero, O.A.; Frontana-Uribe, B.A. 2018. Conducting polymers in the fields of energy, environmental remediation, and chemical–chiral sensors. *Chemical Reviews* 118:4731–4816.
21. Bisquert, J. 2008. The two sides of solar energy. *Nature Photonics* 2:648–649.
22. Wu, J.; Li, Y.; Tang, Q.; Yue, G.; Lin, J.; Huang, M.; Meng, L. 2014. Bifacial dye-sensitized solar cells: A strategy to enhance overall efficiency based on transparent polyaniline electrode. *Scientific Reports* 4:4028.
23. Kang, J.S.; Kim, J.; Kim, J.Y.; Lee, M.J.; Kang, J.; Son, Y.J.; Jeong, J.; Park, S.H.; Ko, M.J.; Sung, Y.E. 2018. Highly efficient bifacial dye-sensitized solar cells employing polymeric counter electrodes. *ACS Appl. Mater. Interfaces* 10:8611–8620.
24. Tai, Q.; Chen, B.; Guo, F.; Xu, S.; Hu, H.; Sebo, B.; Zhao, X.Z. 2011. In situ prepared transparent polyaniline electrode and its application in bifacial dye-sensitized solar cells. *ACS Nano* 5:3795–3799.
25. Hecht, D.S.; Hu, L.; Irvin, G. 2011. Emerging transparent electrodes based on thin films of carbon nanotubes, graphene, and metallic nanostructures. *Advanced Materials* 23:1482–1513.
26. Yu, L.; Shearer, C.; Shapter, J. 2016. Recent development of carbon nanotube transparent conductive films. *Chemical Reviews* 116:13413–13453.
27. Ding, S.; Tian, Y. 2019. Recent progress of solution-processed Cu nanowires transparent electrodes and their applications. *RSC Advances* 9:26961–26980.

28. Zhang, Y.; Guo, J.; Xu, D.; Sun, Y.; Yan, F. 2018. Synthesis of ultralong copper nanowires for high-performance flexible transparent conductive electrodes: The effects of polyhydric alcohols. *Langmuir* 34:3884–3893.
29. Guo, H.; Lin, N.; Chen, Y.; Wang, Z.; Xie, Q.; Zheng, T.; Gao, N.; Li, S.; Kang, J.; Cai, D.; Peng D.L. 2013. Copper nanowires as fully transparent conductive electrodes. *Scientific Reports* 3:2323.
30. Ding, S.; Tian, Y.; Jiu, J.; Suganuma, K. 2018. Highly conductive and transparent copper nanowire electrodes on surface coated flexible and heat-sensitive substrates. *RSC Advances* 8:2109–2115.

CHAPTER 10

Superconductors as Energy Materials

10.1 SUPERCONDUCTIVITY AND SUPERCONDUCTORS

One of the most serious problems in almost all typical electric energy provision schemes is resistive losses during, for example, electric power transmission or distribution. Widely used metals have noticeable electrical resistance to direct current (DC) and alternating current (AC) at temperatures close to those typical on the Earth's surface. However, there is a class of materials that exhibit superconducting properties at lower temperatures. These materials can be used to minimize the abovementioned resistive losses, design unique energy storage systems, and enable faster transportation, to name a few areas of their possible applications. Let us start with a general definition of superconductivity.

DEFINITION:

Superconductivity is the phenomenon of materials exhibiting a zero measured electrical resistance to DC.

Three boundary conditions should be "just right" to achieve the superconducting state of a specific material. Each of these criteria has critical values beyond which superconductivity is not observed. These are the temperature, T_c, magnetic field, H_c, and current density, I_c (see Figure 10.1).

In practice, the most important parameter is the critical temperature or, more precisely, transition temperature. Above this, there is usually no obvious indication that the material might be a superconductor. The critical

DOI: 10.1201/9781003025498-10

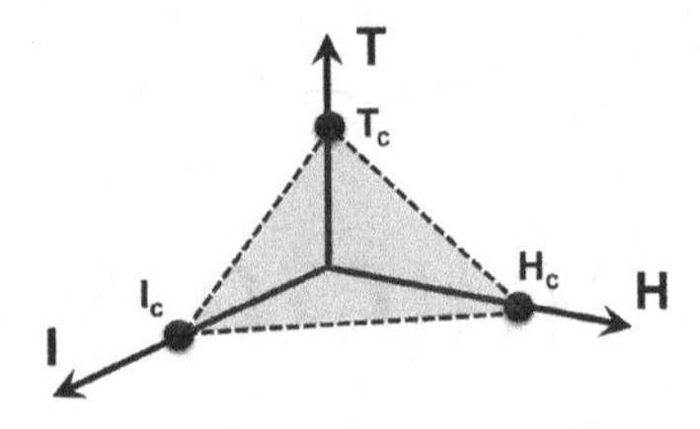

FIGURE 10.1 Schematic depiction of the three boundary conditions, such as critical temperature, T_c, critical DC, I_c, and critical magnetic field, H_c, below which the superconducting state of materials is observed.

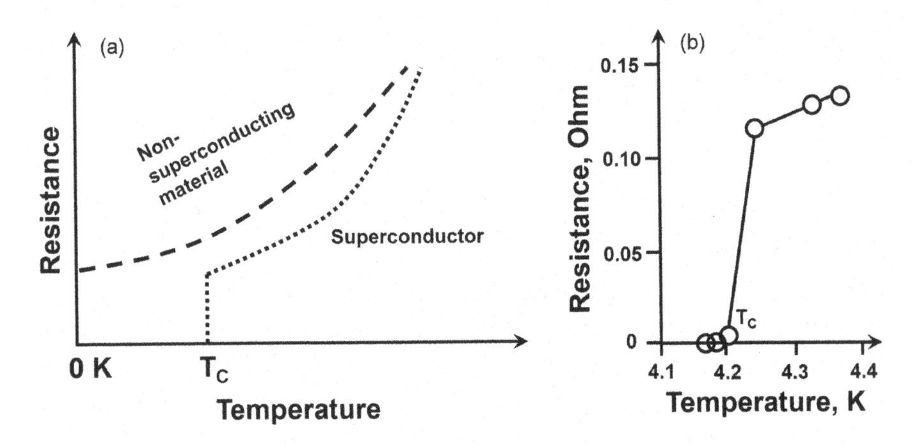

FIGURE 10.2 (a) Approximate dependencies of the electrical resistance for nonsuperconductors and superconducting materials as a function of temperature. (b) Dependence of the resistance of metallic mercury on temperature. ((b) is adapted from [1].)

temperature for superconductors is the one at which the electrical resistance of a material abruptly drops to basically zero values (Figure 10.2). It is also important to note that the superconducting state cannot exist in the presence of a magnetic field or current greater than their critical values, *even at close to absolute zero temperature.*

For the first time, superconductivity was observed for metallic mercury, Hg, by Heike Kamerlingh Onnes (the Nobel Prize in physics 1913 *for his investigations on the properties of matter at low temperatures, which led, inter alia, to the production of liquid helium*) and his doctoral student Gilles Holst in 1911 [1,2]. For many years afterward, superconductivity was thought to consist simply of the disappearing of all electrical resistance below some transition temperature. Interestingly, it was not a surprising effect at that time. This rather simple way of thinking was dominating

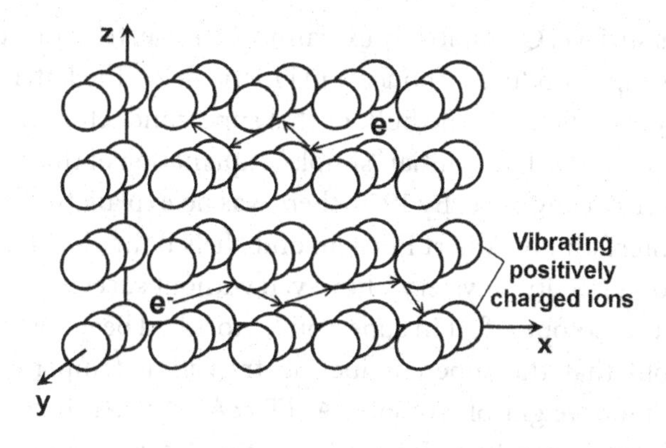

FIGURE 10.3 The Drude model tries to apply the classic kinetic theory to the microscopic behavior of electrons in a solid. In this model, electrons are constantly rebounding from relatively immobile positive ions (open circles in the Figure) in a crystal. The model can explain Ohm's law semi-quantitatively.

mostly because of the so-called Drude model (1900, see Figure 10.3), which was frequently used at the beginning of the 20th century. According to that model, metal conductivity, σ, can be expressed as follows:

$$\sigma = n \cdot e_0^2 \cdot \tau / m$$

where n is the electron concentration, e_0 is the elementary charge, τ is the relaxation time, and m is the electron effective mass. The main question was about the dependence of the relaxation time on the temperature τ(T). It was expected that at some temperature, the energy of electrons would not be dissipated, e.g., due to the motions of the atoms in the crystal lattice. The only curiosity was to observe such a phenomenon experimentally.

In 1908, liquid helium ($T_{boiling, He} = 4.2$ K) was obtained for the first time, and it was possible to measure the conductivity of pure mercury at temperatures close to or even lower than 4 K. As one can see from Figure 10.2b, indeed, the resistance suddenly drops to zero at *ca* 4.18 K. Liquid helium was not available anywhere in the world outside the Onnes's laboratory until 1923. Therefore, it is not a surprise that before 1928, all superconductors were discovered in Leiden.

The significant advance on the way to understanding superconductivity was the discovery of the Meissner effect (1933), which surprisingly showed that superconductors should also be perfect diamagnets. In other words, the magnetic field is excluded from the bulk of a superconductor below T_c. At the same time, it is not excluded from a relatively thin region near the sample's surface. Qualitatively explaining, the electric currents near the surface of a superconductor induce a magnetic field. The latter cancels the applied magnetic fields in the bulk of this superconductor. This discovery was a point after which the initial "simple" explanation of this phenomenon could not be used anymore. By 1950, there was no explanation of superconductivity. Interestingly, Albert Einstein and Niels Bohr also tried to resolve this superconductivity "mystery", however, without success.

In 1950, the isotope effect in superconductors had been discovered [3,4]. It turned out that the superconducting transition temperatures varied with the atomic weight of isotopes, A, ($T_c \propto A^{-0.5}$) when investigating the isotopes of Hg (Figure 10.4). It was the first and direct evidence for interactions between the electrons and the lattice and that superconductivity should not have just a purely electronic origin. It was therefore recognized that phonons also play an essential role in superconductivity effects.

Based on this experimental finding and previous theoretical concepts by Cooper [6], John Bardeen, Leon Cooper, and Robert Schrieffer developed the first theory (Bardeen-Cooper-Schrieffer (BCS) theory) [7] of superconductivity (Nobel prize 1972).

The central pillar of the BCS theory is that superconductivity can be explained through electrons, which can be attracted to each other

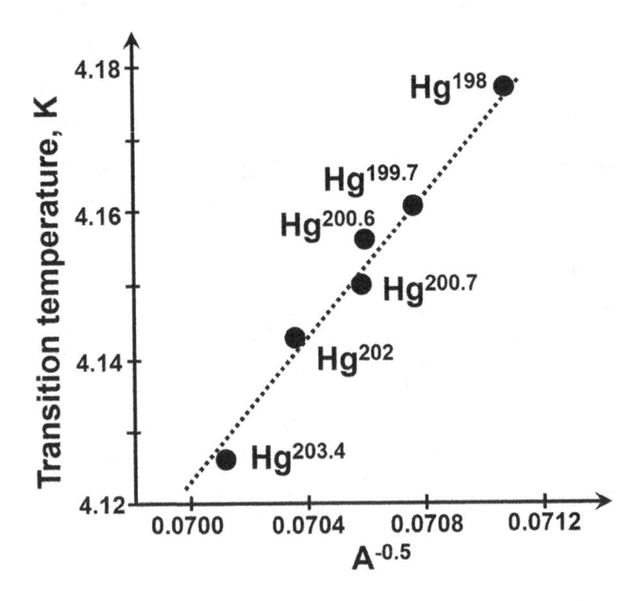

FIGURE 10.4 The dependence of the transition temperature on $A^{-0.5}$ for different isotopes of mercury, where A is the atomic weight of Hg-isotopes. (Adapted from [5]. Data are from [3,4].)

through interaction with the crystal lattice. These electrons form a sort of pairs called *Cooper pairs.* A Cooper pair should be considered as a purely quantum effect, and it is difficult to describe it correctly using macroscopic approaches. However, it can be reasonably represented as a pair of electrons bound together due to the electron-phonon interactions (Figure 10.5). It is noteworthy that these electrons forming a pair are not necessarily close to each other.

For the explanation of the superconductivity, it is essential to note that the Cooper pairs behave like bosons. This means that two or more identical quasi-bosons (Cooper pairs) can occupy the same quantum state. In other words, at some temperature regions, phonons "force" the electrons to become quasi-bosons to lower the overall energy state. This energy lowering is done through the Cooper-pair mechanism. Since the Pauli exclusion principle does not apply for the Cooper pairs anymore, all of them can "drop" into one quantum state. All electrons contributing to the conductivity can therefore have the same energy level. This situation can happen in certain materials below the critical temperature, T_c, and the other critical parameters such as the current and magnetic field. Therefore, the electrons of an entire superconducting specimen can exhibit behavior analogous to that of a single atom or molecule. The superconducting state can be consequently referred to as a unique *macroscopic quantum state.*

The electron-phonon interaction results in an energy gap (E_g) between the normal and superconducting phases at the Fermi level (see Figure 10.6).

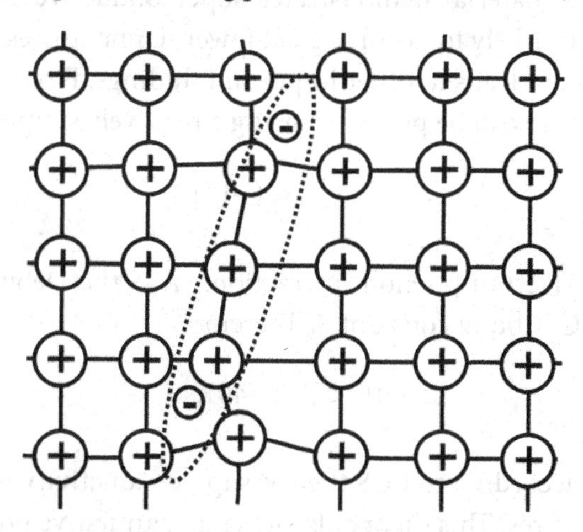

FIGURE 10.5 Schematics of a phonon-mediated formation of a Cooper pair. In BCS theory, the potential energy of the Cooper pairs is lowered through their interactions with the phonons.

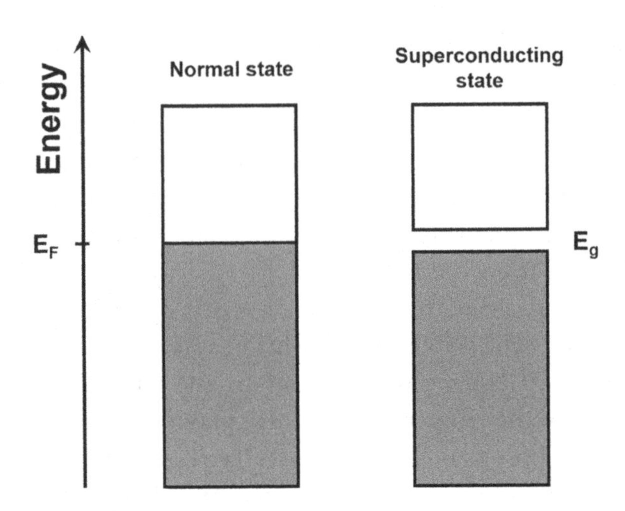

FIGURE 10.6 Schematic energy diagrams of metal in normal and superconducting states. To convert the superconducting metal back to its normal state, one should overcome an energy barrier, E_g, which is proportional to $3.5 \cdot k \cdot T_c$ at 0 K according to BCS theory.

Recalling the mechanism of how the active resistance works, one can conclude that in order for one electron to change its energy in a superconducting phase, it is necessary to destroy the whole macroscopic quantum state so that all electrons should become fermions again (one has to overcome the energy barrier E_g). One either destroys the macroscopic quantum state, or a material demonstrates superconductive properties. The destruction is unlikely to occur, e.g., at lower temperatures.

The BCS theory leads to other important findings. First of all, the transition temperature can be predicted using a relatively simple equation:

$$T_c = A_D \cdot \exp(-1/B)$$

where for the electron-phonon interactions, $A_D \sim$ the Debye temperature and $B \ll 1$ (a BCS "bond constant"). Therefore:

$$0\,K < T_c <\sim 40\,K$$

meaning that according to BCS theory, superconductivity is only possible at low temperatures. This theory also gives a quantitative prediction of the energy gap, E_g, at 0 K as

$$E_g = 3.5 \cdot kT_c$$

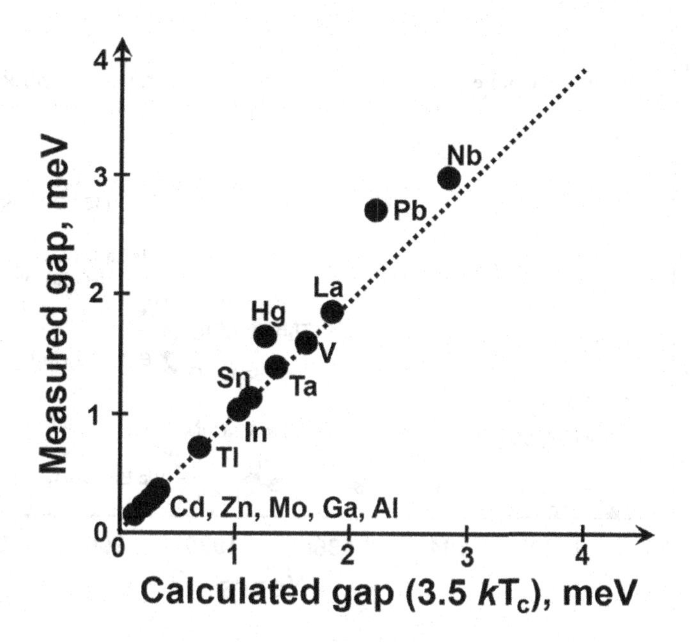

FIGURE 10.7 Energy gaps for some superconducting metals plotted as measured vs. calculated using BSC theory values. (Adapted from [5].)

Figure 10.7 shows that there is an excellent correlation between the measured and BCS-predicted energy gap for a number of superconducting metals.

However, limitations of BCS theory became obvious in 1986–1987, when a new class of ceramic copper-oxide-based superconductors (cuprates) was discovered [8–10]. The state-of-the-art cuprates found afterward could carry the currents without losing energy as heat at temperatures of up to 164 K or −109°C (see Figure 10.8). According to the BCS theory of conventional low-temperature superconductors, these copper oxides would have seemed the least likely materials in which to look for such properties. At room temperature, they are relatively poor conductors. The unexpected discovery of *high-temperature* superconductivity in the copper oxides triggered a massive amount of innovative scientific research. Unfortunately, up to now, no quantitative theory describing high-temperature superconductivity is widely accepted, and the discovery of new superconductors is rather an art [11]. However, there are hypotheses that not only phonon-electron interactions are responsible for the superconductive properties.

Despite certain challenges in elaborating new theoretical concepts capable of explaining and predicting the high-temperature superconductivity, within a few decades after its discovery, new benchmarks in

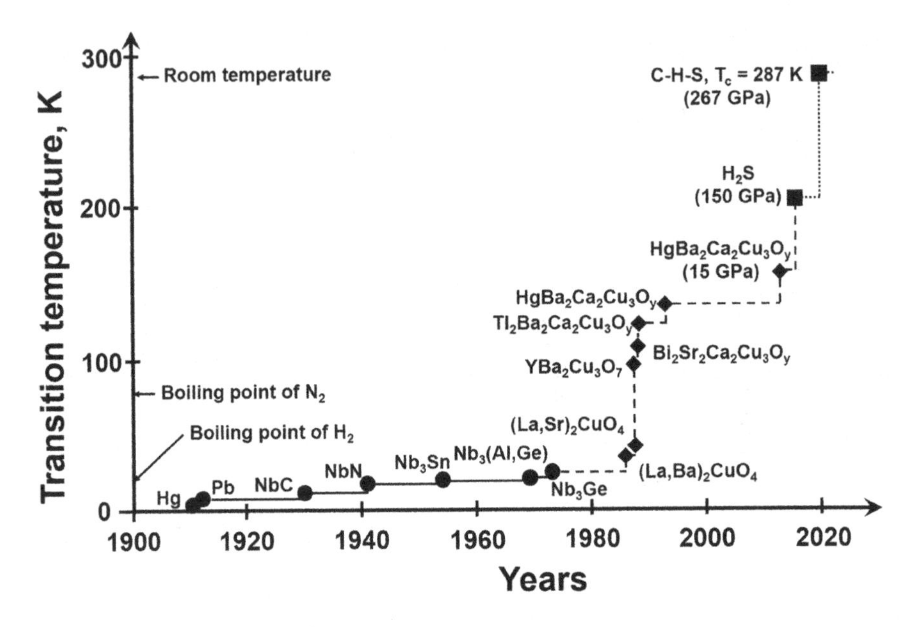

FIGURE 10.8 The development of superconductors with time toward materials with the transition temperature, T_c, closer to room temperature. Important "milestones" are indicated.

the transition temperatures were established under high-pressure conditions (see Figure 10.8). For example, H_2S at 150 GPa pressure showed T_c over 200 K.

One should also mention the discovery of superconductivity in magnesium diboride, MgB_2, below 39 K in 2001 [12]. This superconductor is based on abundant raw materials and enables liquid hydrogen for cooling. Room temperature superconductivity at high pressures was discovered in 2020 [13]. It was found in specific C-H-S systems (the so-called carbonaceous sulfur hydrides). Figure 10.9 illustrates that a superconducting transition at ~287 K is observed at the pressure of *ca* 267 GPa. What is also remarkable is that new computational methods together with semi-empirical approaches may allow discovering even higher-temperature (with T_c close to 473 K) superconducting materials [14].

Although thousands of superconductors have been identified by now, only a few of them are currently considered for energy applications. Those are Nb-Ti (fully commercialized), Nb_3Sn (fully commercialized), copper-based oxide (early stage of commercialization), MgB_2 (early stage of commercialization), and iron-based [15,16] (laboratory research) [17] superconductors.

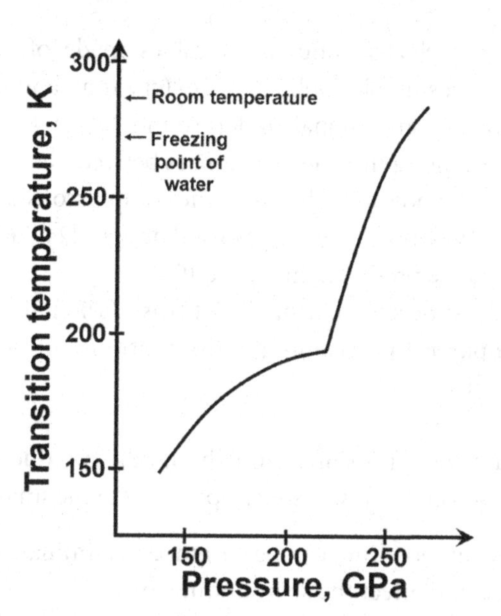

FIGURE 10.9 Transition temperature, T_c, as a function of the applied pressure in the carbonaceous sulfur hydride system. A superconducting transition at ~287 K is observed at the highest pressure measured, ca 267 GPa. (Adapted from [13].)

10.2 SUPERCONDUCTORS FOR TRANSMISSION LINES

Probably, the most straightforward areas of energy applications of superconductors are the electric transmission lines to make a viable alternative to the existing overhead lines [18–20]. Moreover, if large "green-power" resources are widely employed in remote places like deserts to serve growing load centers, innovative methods of transmitting GW-level power over long distances to the customers will often be required [21]. One way to achieve this is to use DC cables based on high-temperature or even low-temperature superconductors [22,23]. With the current state-of-the-art superconducting materials being used in transmission lines, they seem viable if the "green" energy should be transmitted to distances of thousands of kilometers, potentially saving metals, space, and installation costs. In addition to the benefits of minimizing the resistive losses, one can also primarily address the common AC/DC issue. The latter originates from the fact that most of the common electrical devices in households and industries are intrinsically DC devices. If one uses no AC/DC converters and utilizes only DC superconducting electric transmission lines, it is possible to lower the final electricity costs even further [24].

Superconducting electric lines need cables made of superconducting materials placed in a suitable cooling system (cryostat) and mounted underground similarly to conventional underground cables. Cryogenic stations can be constructed along the line at distances between 10 and 100 km, comparably to those used nowadays between the compressor stations for natural gas long-distance transport pipelines. Several designs [24–26] were suggested with a general scheme presented in Figure 10.10.

Long-distance superconducting DC transmission lines have serious advantages compared to conventional installations, and the most important of them are listed below.

- They would have no impact on the landscape due to their underground location, i.e., low visual impact due to the small size is possible
- These transmission lines can be designed to eliminate electromagnetic fields that could affect the surrounding area
- They would have a smaller environmental footprint than both overhead lines and standard underground cables
- These lines would minimize the land use and property acquisition, leaving the value of local real estate unaffected

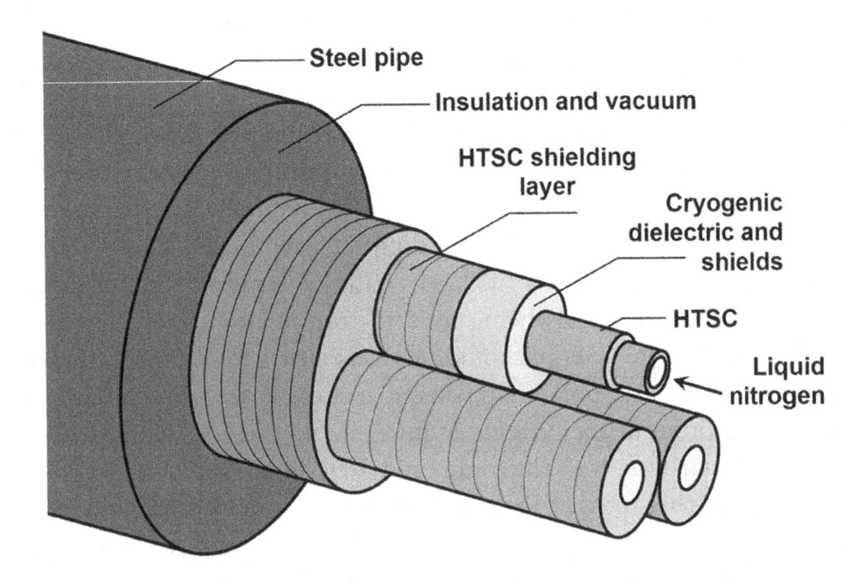

FIGURE 10.10 A simplified structure of a long-distance DC transmission cable. HTSC – high-temperature superconductor.

- They would not be influenced by natural weather phenomena such as wind, fog, snow, or ice
- They typically have a much higher effective current density of *ca* 100 A/mm^2 if compared with copper cables (1–5 A/mm^2)

One can distinguish the following milestones on the way to broader use of superconducting transmission electric lines, as listed below.

1967: Evaluation of the possibility of transferring 100 GW over 1000 miles in a single superconducting DC power cable in the US (Nb_3Sn, 4 K, liquid helium)

2001: The Electric Power Research Institute introduced a concept for large-scale energy transfer with liquid H_2 using MgB_2 superconductors

2008: Long Island Power Authority, New York, US; a 600-meter-long underground cable was installed connecting a power station to the overhead line network

2009: The technology is ready for commercial applications with liquid N_2 using high-temperature superconductors

2014: AmpaCity [27] system installation in Essen, Germany (the world-wide longest superconducting power cable in the city by now) and South Korea's pilot project to build the first "real-world" distribution system based on superconducting cables

2021: Tests of full-scale HTS transmission cable line (2.4 Km), St. Petersburg, RF [28]

At the same time, it is important to mention the main impediments, which currently slow down the commercial application of such technologies in electrical energy transmission.

- There is a misbalance between either the high costs of the high-temperature superconductors or expenses for the cooling gas/systems
- Complications to repair the cable damages. It is now a relatively time-consuming process
- Losses when using AC. There is no generally accepted universal physical model to describe these losses

- Thermal leaks between the cold liquid nitrogen and the warm surroundings
- The strength of the current, which one can use in these cables, is limited. At some I_c, the superconducting state will be lost
- Maintenance issues are a big question. This still should be extensively tested in pilot projects

10.3 SUPERCONDUCTORS FOR ENERGY STORAGE SYSTEMS AND TRANSPORTATION

The basic idea behind the use of superconductors in energy storage devices is relatively simple. It is based on constructing and using induction coils made of superconductors and cooled below the critical temperature (Figure 10.11) [29–31]. Energy stored in a standard inductor will fade out rather quickly due to the ohmic resistance in the coil when the power supply is disconnected. The time constant of a coil is $\tau = L/R$. If R goes to zero in the case of superconductors, then τ goes to infinity, enabling theoretically unlimited time to store electric energy in the magnetic field.

The magnetic energy, E, stored is calculated according to the following equation connecting the inductance, L, and the current flowing, I:

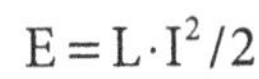

$$E = L \cdot I^2/2$$

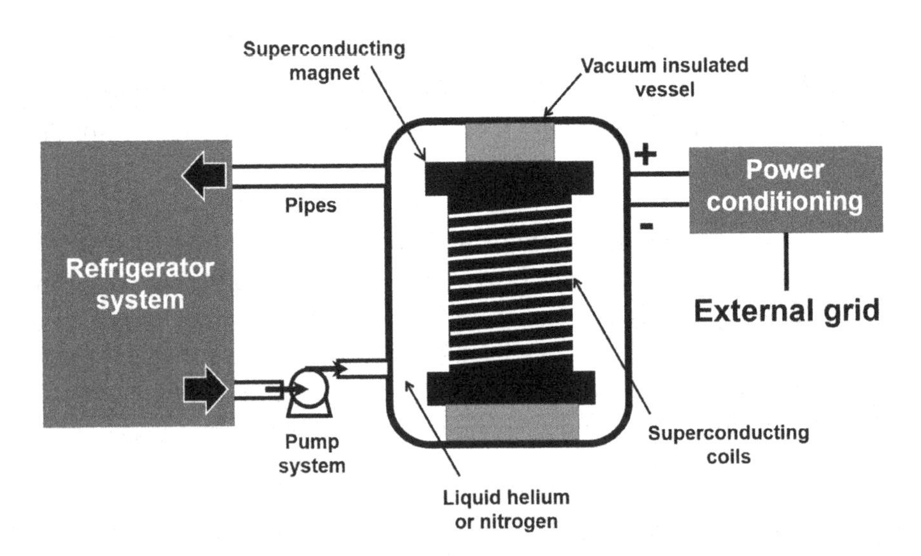

FIGURE 10.11 A schematic diagram of a SMES system. (Adapted from [32].)

As discussed above, the conductor has to be operated below a critical temperature T_c, below a critical current I_c, and below a critical magnetic field H_c. Superconducting magnetic energy storage (SMES) systems store energy within the magnetic field created by the flow of DC in a superconducting coil with near-zero loss of energy. A typical SMES consists of two parts: a cooled superconducting coil (e.g., Nb/Ti or Y-Ba-Cu-O ceramics) and a power conditioning system (Figure 10.11). Ideally, once the superconducting coil is charged, the current will not decay, and the magnetic energy can be stored indefinitely long. The energy will be stored until the coil is again connected to the grid to be discharged. The conversion efficiency in one cycle using such systems is in practice *ca* 95%–97%.

Relatively large systems are already in use by power generation and transmission utilities to help to stabilize long transmission lines and provide a very fast power backup in times of crisis or peak loads.

The main advantages of SMES systems are listed below (see also Table 10.1).

1. The short time between the charging and discharging (almost instantaneous)
2. The very low energy loss in the charging and discharging process
3. The absence of moving parts in the system's main components

The largest drawback of the state-of-the-art SMESs so far is their price. One additional disadvantage regarding, e.g., high-temperature ceramic superconductors is the fact that the materials being used are brittle and hard to shape, at least into the form of a coil.

TABLE 10.1 Comparison of the Performance of Some Energy Storage Systems

	Typical Range/Response Time	Life Time/Backup Time
SMES	1–100 MW/milliseconds	~30 years/seconds
Supercapacitors	1–250 kW/milliseconds	~10–20 years/seconds
Compressed air energy storage system	25–350 MW/1–2 min	More than 50 years/hours
Flywheel energy storage system	~kW/1–2 min	~20 years/minutes
Pumped hydro energy storage system	Up to ~2 GW	~50 years/days
Batteries	~20 MW/seconds	~3–7 years/hours

Relatively affordable commercial systems are available in the market for energy applications. SMESs are indeed used now to store energy, e.g., D-SMES in Wisconsin, USA 3MW units, or in Kameyama (Japan). South Korea has been actively developing HTSMES since 2004 to be widely used in commercial grids. There were two national projects, one for 600 kJ and the other for 2.5 MJ SMES.

Superconductors attract significant attention for public transportation, trains, or trams [33]. These trains use the effect of magnetic levitation (the so-called *maglev* technology). There is an approximately 10 mm gap between the *maglev* transport mean and the track minimizing losses due to friction. The superconducting magnet material is normally placed on board, while the role of the track is to be a guideway. There are currently two major operational lines. One line is a relatively short commercial one from Shanghai airport, and another one is the more-than-40-km test track in Japan. Other prototypes have been recently demonstrated, or plans were announced by companies in Germany, China, and other countries. The expected and demonstrated speeds of these trains are more than 500 km/h. This is a competitive speed compared to airplanes and reduces the CO_2 emission by at least a factor of two.

10.4 SUMMARY AND CONCLUSIONS

Wider application of superconducting materials for energy provision and storage would noticeably change everyday life. First of all, this should decrease the overall resistive energy losses in transmitting high-level power over long distances and enable faster and better long-range passenger transportation. In addition, superconducting magnetic storage systems should complement the electrical energy storage device portfolio to balance energy provision at peak demands. However, the current understanding of superconducting materials impedes the rational development of affordable high-temperature superconductors necessary for their broader commercial applications in energy provision.

The well-accepted BCS theory can predict properties of the superconductors with the transition temperature, T_c, below 40 K. It has limitations to explain the properties of high-temperature superconductors. The latter are now discovered mainly using semi-empirical approaches. Nevertheless, state-of-the-art high-temperature superconductors, the so-called cuprates, are currently used, and liquid nitrogen is typically utilized to maintain their superconducting properties. What is more, a few classes of compounds demonstrate the superconducting transition temperature close to the room temperature under high-pressure conditions. Some hydrides are predicted even to exhibit superconductivity at elevated temperatures. However, further development of the theory of superconductivity is necessary to enable faster progress in this field of energy materials.

Current applications of superconductors are not limited to energy provision schemes. They are also used in some transportation systems and are widely applicable in devices for medicine.

10.5 QUESTIONS

1. Define superconductivity.
2. What are the three main criteria (boundary conditions) for the material to demonstrate superconductive properties?
3. Describe the phenomenology of the Meissner and isotope effects in superconducting materials?
4. Why did the Meissner effect and isotope effect contribute to the understanding of superconductivity?
5. Explain the origin of superconductivity using BCS theory.
6. Name the most important classes of high-temperature superconductors.
7. Does BCS theory explain the mechanism of superconductivity in high-temperature superconductors (and why)?
8. Name state-of-the-art superconducting materials for energy provision and storage.
9. What are the basic concepts to use high- and low-temperature superconductors for electrical energy transmission, energy storage systems, and transportation? Analyze the *pros* and *cons* of these concepts.

REFERENCES

1. Onnes, H.K. 1911. The resistance of pure mercury at helium temperatures. *Communications from the Laboratory of Physics at the University of Leiden* 12:120.
2. van Delft, D.; Kes, P. 2010. The discovery of superconductivity. *Physics Today* 63:38–43.
3. Maxwell, E. 1950. Isotope effect in the superconductivity of mercury. *Physical Review* 78:477.
4. Reynolds, C.A.; Serin, B.; Wright, W.H.; Nesbitt, L.B. 1950. Superconductivity of isotopes of mercury. *Physical Review* 78:487.

5. Rohlf, J.W. 1994. *Modern Physics from α to Z^0*. John Wiley & Sons, Inc.: Hoboken (chapter 15), p. 417.
6. Cooper, L.N. 1956. Bound electron pairs in a degenerate Fermi gas. *Physical Review* 104:1189–1190.
7. Bardeen, J.; Cooper, L.; Schriffer, J.R. 1957. Theory of superconductivity. *Physical Review* 108:1175–1204.
8. Bednorz, J.G.; Müller, K.A. 1986. Possible high T_c superconductivity in the Ba–La–Cu–O system. *Zeitschrift für Physik B Condensed Matter* 64:189–193.
9. Wu, M.K.; Ashburn, J.R.; Torng, C.J.; Hor, P.H.; Meng, R.L.; Gao, L.; Huang, Z.J.; Wang, Y.Q.; Chu, C.W. 1987. Superconductivity at 93 K in a new mixed-phase Y-Ba-Cu-O compound system at ambient pressure. *Physical Review Letters* 58:908–910.
10. Michel, C.; Hervieu, M.; Borel, M.M.; Grandin, A.; Deslandes, F.; Provost, J.; Raveau, B. 1987. Superconductivity in the Bi-Sr-Cu-O system. *Zeitschrift für Physik B Condensed Matter* 68:421–423.
11. Gui, X.; Lv, B.; Xie, W.W. 2021. Chemistry in superconductors. *Chemical Reviews* 121:2966–2991.
12. Nagamatsu, J.; Nakagawa, N.; Muranaka, T.; Zenitani, Y.; Akimitsu, J. 2001. Superconductivity at 39 K in magnesium diboride. *Nature* 410:63–64.
13. Snider, E.; Dasenbrock-Gammon, N.; McBride, R.; Debessai, M.; Vindana, H.; Vencatasamy, K.; Lawler, K.V.; Salamat, A.; Dias, R.P. 2020. Room-temperature superconductivity in a carbonaceous sulfur hydride. *Nature* 586:373–377.
14. Sun, Y.; Lv, J.; Xie, Y.; Liu, H.; Ma, Y. 2019. Route to a superconducting phase above room temperature in electron-doped hydride compounds under high pressure. *Physical Review Letters* 123:097001.
15. Kamihara, Y.; Watanabe, T.; Hirano, M.; Hosono, H. 2008. Iron-based layered superconductor $La[O_{1-x}F_x]FeAs$ ($x=0.05–0.12$) with $T_c=26$ K. *Journal of the American Chemical Society* 130:3296–3297.
16. Hosono, H.; Yamamoto, A.; Hiramatsu, H.; Ma, Y. 2018. Recent advances in iron-based superconductors toward applications. *Materials Today* 21: 278–302.
17. Yao, C.; Ma, Y. 2021. Superconducting materials: Challenges and opportunities for large-scale applications. *iScience* 24:102541.
18. Doukas, D.I. 2019. Superconducting transmission systems: review, classification, and technology readiness assessment. *IEEE Transactions on Applied Superconductivity* 29:5401205.
19. Thomas, H.; Marian, A.; Chervyakov, A.; Stückrad, S.; Salmieri, D.; Rubbia, C. 2016. Superconducting transmission lines – Sustainable electric energy transfer with higher public acceptance? *Renewable and Sustainable Energy Reviews* 55:59–72.
20. Morandi, A. 2015. HTS DC transmission and distribution: Concepts, applications and benefits. *Superconductor Science and Technology* 28:123001.
21. Semeraro M.A. 2021. Renewable energy transport via hydrogen pipelines and HVDC transmission lines. *Energy Strategy Reviews* 35:100658.

22. Chen, Y.; Jiang, S.; Chen, X.Y.; Wang, Y.F.; Li, T. 2020. Preliminary design and evaluation of large-diameter superconducting cable toward GW-class hybrid energy transfer of electricity, liquefied natural gas, and liquefied nitrogen. *Energy Science & Engineering* 8:1811–1823.
23. Jin, J.X.; Chen, X.Y.; Qu, R.; Fang, H.Y.; Xin, Y. 2015. An integrated low-voltage rated HTS DC power system with multifunctions to suit smart grids. *Physica C: Superconductivity and Its Applications* 510:48–53.
24. Thomas, H.; Marian, A.; Chervyakov, A.; Stückrad, S.; Salmieri, D.; Rubbia, C. Superconducting transmission lines – Sustainable electric energy transfer with higher public acceptance? *Renewable and Sustainable Energy Reviews* 55:59–72.
25. Cheetham, P.; Viquez, J.; Kim, W.; Graber, L.; Kim, C.H.; Pamidi, S.V. 2018. High-temperature superconducting cable design based on individual insulated conductors. Advances in *Materials Science and Engineering.* 2018:3637873.
26. Chen, Y.; Jiang, S.; Chen, X.Y.; Wang, Y.F.; Li, T. 2020. Preliminary design and evaluation of large-diameter superconducting cable toward GW-class hybrid energy transfer of electricity, liquefied natural gas, and liquefied nitrogen. *Energy Science & Engineering* 8:1811–1823.
27. Herzog, F.; Kutz, T.; Stemmle, M.; Kugel, T. 2016. Cooling unit for the AmpaCity project – One year successful operation. *Cryogenics* 80:204–209.
28. Sytnikov, V.; Kashcheev, A.; Dubinin, M.; Karpov, V.; Ryabin, T. 2021. Test results of the full-Scale HTS transmission cable line (2.4 Km) for the St. Petersburg project. *IEEE Transactions on Applied Superconductivity* 31:5400805.
29. Mukherjee, P.; Rao, V.V. 2019. Design and development of high temperature superconducting magnetic energy storage for power applications - A review. *Physica C: Superconductivity and its Applications* 563:67–73.
30. Breeze, P. 2018. Chapter 5- Superconducting magnetic energy storage. In: *Power System Energy Storage Technologies.* Academic Press, Elsevier, Cambridge, MA, pp. 47–52.
31. Tixador, P. 2012. Superconducting magnetic energy storage (SMES) systems. *Woodhead Publishing Series in Energy* 9:294–319. Woodhead Publishing Limited.
32. Luo, X.; Wang, J.; Dooner, M.; Clarke, J. 2015. Overview of current development in electrical energy storage technologies and the application potential in power system operation. *Applied Energy* 137:511–536.
33. Nishijima, S.; Eckroad, S.; Marian, A.; Choi, K.; Kim, W.S.; Terai, M.; Deng, Z.; Zheng, J.; Wang, J.; Umemoto, K.; Du, J.; Febvre, P.; Keenan, S.; Mukhanov, O.; Cooley, L.D.; Foley, C.P.; Hassenzahl, W.V.; Izumi, M. 2013. Superconductivity and the environment: A roadmap. *Superconductor Science and Technology* 26:113001.

CHAPTER 11

Permanent Magnets for Motors and Generators

11.1 INTRODUCTION

In order to maintain the growth of the "green"-energy economy, one needs efficient and affordable magnetic materials, particularly for electric generators and motors [1]. For example, these materials are required for wind turbine electrical generators. In 2020, the globally installed wind generators gave more than 600 GW_p (where subscript p designates the peak power), and the number of such generators is likely to grow substantially, leading to an increased demand for their components. However, there is an important issue. The wind turbines, or the so-called "windmills", have to operate at very low revolution speeds. Typical values range from a minimum of 5 rotations per minute (rpm) to a maximum of 20 rpm. Because of this, plus peculiarities of current technologies in this field, one has to use in the majority of cases *permanent magnets* (PMs). PMs are advantageous in wind power production at a low rotation speed. On average, medium-speed geared PM-wind turbine generators need ~200 kg of magnetic materials per MW_p [2].

Furthermore, if hydrogen is utilized in automotive applications as a fuel, electrical motors are necessary. Electrical motors are also needed if different types of batteries or supercapacitors are to be widely used in cars, buses, trains, and other means of transportation. Most electric vehicles designed for transportation purposes currently use PM motors [3]:

DOI: 10.1201/9781003025498-11

they are approximately 70% smaller than induction motors (IMs). At the same time, the IMs are approximately 2.5 times heavier than the PM-based vehicles.

It is surprising that while our civilization knows PMs for more than 2000 years, one still experiences a noticeable deficit of these kinds of materials, especially for the current energy applications. This issue should be addressed, and one can only agree that there is *an existing but somehow largely ignored fact that improved magnetic materials are a solution to the energy crisis. The solution via development in magnetic materials will give an immediate impact in producing renewable energy and in raising energy efficiency* [4].

The story of understanding and development of PMs probably begins with a mineral called magnetite (Fe_3O_4), one of the first natural magnetic materials found approximately 2500 years ago in the district of Magnesia in the present day Turkey. The first well-received scientific study of magnetism was made by William Gilbert, who published his classical book "On the Magnet" in 1600. In the early 19th century, Hans Christian Oersted discovered that an electric current induces a magnetic field. This event was a real breakthrough in the understanding of electricity itself. In 1907, Pierre-Ernest Weiss supposed that in addition to any externally applied magnetic field, H, there should be a n internal "molecular" field in a (ferro)magnet proportional to its magnetization. Later, Pierre Curie discovered that certain magnetic materials undergo a sharp change in their magnetic properties at a transition temperature, which is now called Curie temperature.

The discovery of a new family of materials, namely the ferrimagnetic hexagonal ferrites (e.g., $SrFe_{12}O_{19}$) in 1951 [5], for which the shape barrier was broken, designated a new milestone in the development of PMs. For those materials, self-demagnetization did not quickly occur due to the shape issues, and bar or horseshoe shape became possible.

In 1982, General Motors and Sumitomo Special Metals independently discovered the $Nd_2Fe_{14}B$ compound [6], a magnetic material that, for example, can easily lift thousands of times its own weight. The energy product $|BH|_{max}$ (see the following sections for explanations) of magnetic materials improved quasi-exponentially during the 20th century, doubling approximately every 12 years. However, it has not improved considerably in the last ~20 years (see Figure 11.1). In the following, some key theoretical aspects related to the understanding and design of magnetic materials are presented.

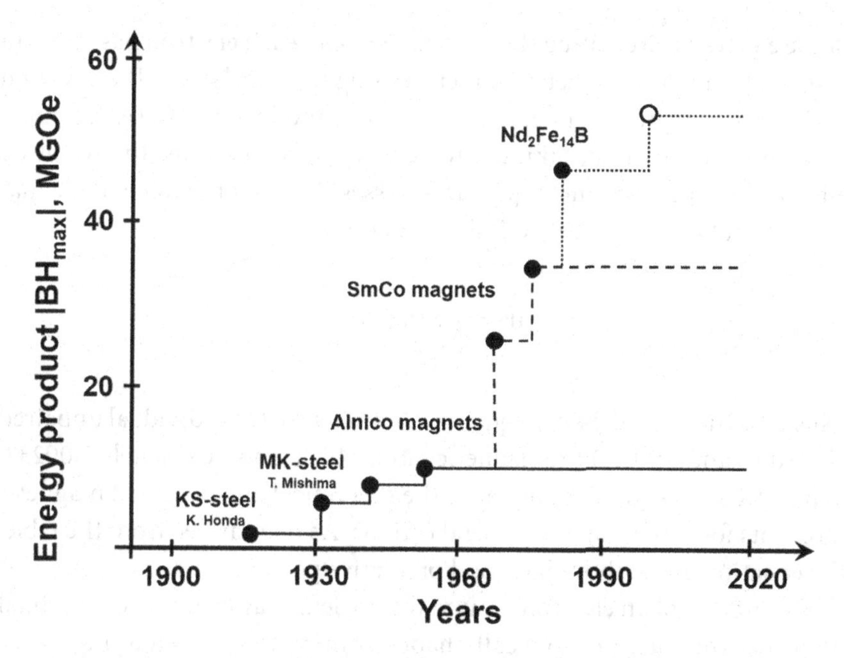

FIGURE 11.1 Development of PM materials with time. The energy product, one of the central figures of merit in the field of magnetic materials, is expressed in Mega-Gauss-Oersted, 1MGOe=7957.75 J/m^3. (The data are taken from [7,8].)

11.2 THEORETICAL CONSIDERATIONS

The theoretical description of PMs can be performed using two historically established ways. The first one can be arbitrarily named "physics way" in terms of circulating currents. In this way of thinking, one can envisage magnetism as a sort of side effect of electron motion. The second one can analogously be called "engineer way", which describes magnetism in terms of magnetic poles. The latter can be found, for example, at the ends of a bar magnet.

It is known that inorganic solids that exhibit magnetic effects dissimilar to diamagnetism, a specific property of all substances, have constituent parts with some unpaired electrons in their outer valence shells. In particular, unpaired electrons are usually "localized" on metal cations. The unpaired electrons can have both spin and orbital motion, generating a magnetic moment associated with the electrons. Magnetic behavior is restricted mainly to compounds of transition metals and lanthanoids, most of which possess unpaired d and f electrons, respectively.

The magnetic moment of a free atom in the absence of a magnetic field consists of two contributions. The first one is the orbital angular momenta of the electrons circulating the nucleus. Second, each electron has an extra contribution to its magnetic moment arising from its "spin". The spin and orbital angular momenta can combine and produce the observed magnetic moment. At first, taking into account only the spins, the resulting effective magnetic moment, μ_S, can be assessed using the following equation (in units of the so-called Bohr magnetron):

$$\mu_S = g\left(S(S+1)\right)^{0.5}$$

where S is the sum of the spin quantum numbers of the individual unpaired electrons, and g is the gyromagnetic ratio, which is approximately 2.00232. Table 11.1 shows some examples of the theoretical and observed magnetic moments for some transition metal cations. As one can see from the table, the equation given above has good predictive power.

The motion of an electron around the nucleus may give rise to an orbital moment. This situation typically happens in materials containing heavy metal ions such as lanthanoids. The orbital moment then contributes to the overall magnetic moment. In cases where the orbital moment makes its full contribution, the following equation can be used:

$$\mu_{S+L} = \left(g^2 S(S+1) + L(L+1)\right)^{0.5}$$

where L is the orbital angular momentum quantum number for the metal ion.

TABLE 11.1 Theoretical and Measured (in Bohr Magnetron Units) Magnetic Moments, μ_S, for Some Cations of Transition Metals

Ion	Calculated μ_S	Measured μ_S	Number of Unpaired Electrons
V^{4+}	1.73	1.8	1
Cu^{2+}	1.73	1.7-2.2	1
V^{3+}	2.83	2.8	2
Ni^{2+}	2.83	2.8-4.0	2
Cr^{3+}	3.87	3.8	3
Co^{2+}	3.87	4.1-5.2	3 (high spin)
Fe^{2+}	4.9	5.1-5.5	4 (high spin)
Co^{3+}	4.9	5.4	4 (high spin)
Mn^{2+}	5.92	5.9	5 (high spin)
Fe^{3+}	5.92	5.9	5 (high spin)

In practice, the equation given above may sometimes not predict the actual values. This situation happens because the orbital angular momentum is either entirely or partially reduced when the electric fields of the neighboring atoms or ions limit the orbital motion of the electrons. Calculated and measured effective magnetic moments for ions of the rare-earth metals are given in Table 11.2. It is clear from the table that the measured values of the μ_{S+L} are indeed often quite far from those theoretically estimated. Nevertheless, one can say that the predictive power of the approach described above is rather good.

Individual magnetic moments of ions containing unpaired electrons are often randomly oriented in a crystal or, in general, in a medium. Partial alignment occurs on the application of a magnetic field. In some materials (the so-called ferro- and antiferromagnets, which are discussed later), the alignment of magnetic dipoles occurs spontaneously, resulting in a total magnetic moment of a material's volume. In this case, one can assume some type of interaction between neighboring spins that allows this to happen (Figure 11.2).

DEFINITION:

Magnetization, M [Oersted or A/m], is the total magnetic moment per unit volume.

Let us consider in the following some particular cases existing in nature and related to the magnetic behavior of matter.

TABLE 11.2 Theoretical and Measured (in Bohr Magnetron* Units) Magnetic Moments, μ_{S+L}, for Some Cations of Rare-Earth Metals

Ion	Calculated μ_{S+L}	Measured μ_{S+L}	Configuration
Ce^{3+}	2.54	2.4	$4f^1 5s^2 5p^6$
Pr^{3+}	3.58	3.5	$4f^2 5s^2 5p^6$
Nd^{3+}	3.62	3.5	$4f^3 5s^2 5p^6$
Pm^{3+}	2.68	-	$4f^4 5s^2 5p^6$
Sm^{3+}	0.84	1.5	$4f^5 5s^2 5p^6$
Eu^{3+}	0.0	3.4	$4f^6 5s^2 5p^6$
Gd^{3+}	7.94	8.0	$4f^7 5s^2 5p^6$
Tb^{3+}	9.72	9.5	$4f^8 5s^2 5p^6$
Dy^{3+}	10.63	10.6	$4f^9 5s^2 5p^6$
Ho^{3+}	10.6	10.4	$4f^{10} 5s^2 5p^6$
Er^{3+}	9.59	9.5	$4f^{11} 5s^2 5p^6$
Tm^{3+}	7.57	7.3	$4f^{12} 5s^2 5p^6$
Yb^{3+}	4.54	4.5	$4f^{13} 5s^2 5p^6$

*Bohr magneton corresponds to the magnetic moment of a 1s electron in hydrogen.

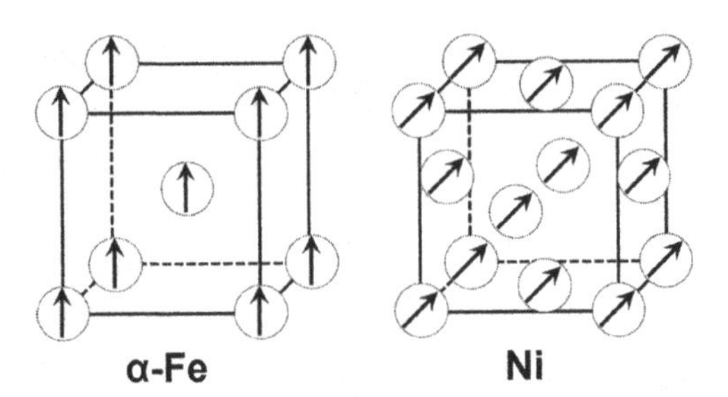

FIGURE 11.2 Alignment of individual magnetic moments in the crystal structure of α-Fe and Ni ferromagnetic materials in the absence of an external field.

One can start with *diamagnetism*, which is one of the fundamental properties of matter originating from the noncooperative behavior of the orbital electrons under the external magnetic fields. The resulting magnetic moment of an atom in a diamagnetic sample is zero. In the external magnetic fields, diamagnetics are magnetized opposite to the applied field direction and, therefore, have negative total magnetization.

In the so-called *paramagnetic* materials, each atom or ion has a certain magnetic moment. However, there is no interaction between these atomic magnets, as shown in Figure 11.3a. In the presence of the external magnetic fields, magnetic moments in this kind of material will be

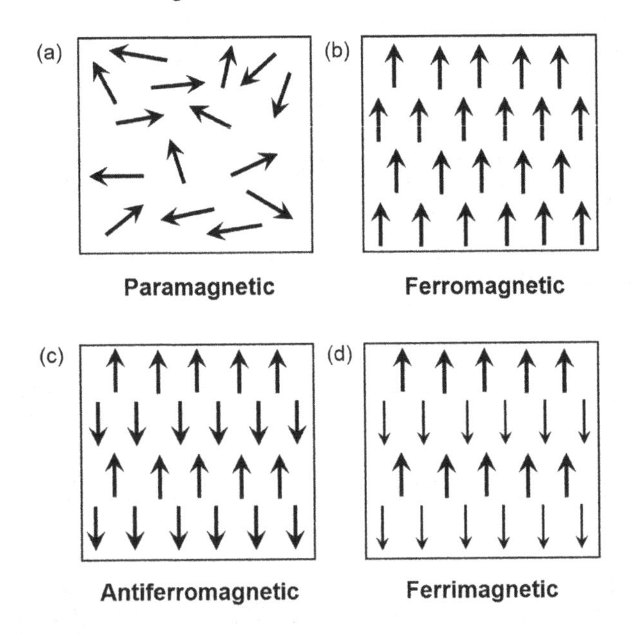

FIGURE 11.3 Types of the ordering of the magnetic dipoles in magnetic materials without an external field.

partially aligned in the direction of the applied field, leading to some net positive magnetization.

In *ferromagnetic* materials (Figure 11.3b), there is a strong interaction between the atomic magnets. The latter are aligned parallel to each other. The specific characteristic of ferromagnetic materials is spontaneous magnetization. It exists in a uniformly magnetized sample in the absence of external magnetic fields. This spontaneous magnetization can be revealed experimentally through *saturation magnetization*: a maximal induced magnetic moment, which an external magnetic field can cause in a specific material.

In *antiferromagnetics*, there are two sublattices with different magnetic moments, which are equal in magnitude and have opposite directions with zero net magnetic moments (Figure 11.3c).

In *ferrimagnetics*, magnetic moments are arranged, similar to the situation with antiferromagnetics. However, the moments are not equal in magnitude; therefore, some spontaneous magnetization can be revealed. The actual arrangements of the magnetic moments are schematically illustrated in Figure 11.3d.

The characteristic feature of ferromagnetic materials is the spontaneous magnetization, M_s, due to a spontaneous alignment of atomic magnetic moments, which disappears upon heating above a critical temperature called the Curie temperature, T_c. Table 11.3 gives some examples of the Curie temperatures for several metals important in designing common magnetic materials.

What is remarkable, when a piece of a ferromagnetic material is cooled below the Curie temperature, the magnetization spontaneously "splits" it into many relatively small regions called *magnetic domains* [9,10]. A magnetic domain can be defined as an area in a magnet in which the magnetization is observed in a uniform direction. In other words, domains

TABLE 11.3 Curie Temperatures for Some Metals

Metal	Curie Temperature, T_c (K)
Er	20
Ho	20
Tm	25
Dy	85
Tb	222
Gd	293
Ni	631
Fe	1043
Co	1404

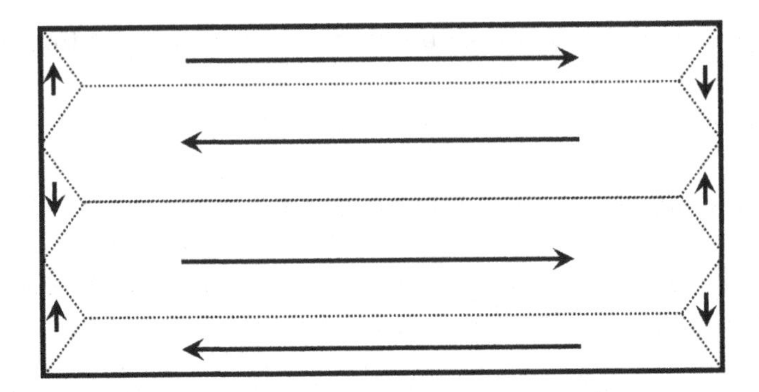

FIGURE 11.4 Schematic two-dimensional representation of a magnetic domain structure of a uniaxial ferromagnet (the Landau–Lifshitz domain configuration).

are considered as groups of spins, which all point to one direction and act cooperatively within the same magnet area. The driving force for such division is the reduction of the overall magnetostatic energy of the system. A schematic representation of these kinds of domains in a piece of material is given in Figure 11.4. One should note that domains significantly influence the design of new powerful magnets, as explained later in this chapter.

11.3 PERMANENT MAGNETS AND THEIR FIGURES OF MERIT

Magnetic materials are categorized as magnetically "soft" and "hard". Soft magnetic materials are those that can be magnetized but rarely stay magnetized for a noticeable time. A typical example, in this case, is metallic iron. Magnetically hard materials, in contrast, can remain magnetized for a long time. PMs, which are necessary for electrical motors and generators, are made from magnetically hard materials. Therefore, in the following, mainly hard magnetic materials will be considered.

In order to understand the central figure of merit for the PMs important in designing the abovementioned motors and generators, one needs to introduce *magnetic induction* defined below.

DEFINITION:

Magnetic induction, B, is the response of a material to a magnetic field, H:

$$B = H + 4\pi M$$

All three parameters introduced here, namely magnetization, M, magnetic induction, B, and magnetic field, H, are used to determine how magnetically *strong* and *hard* a given material is. Let us consider the diagram shown in Figure 11.5a. It schematically illustrates an ideal dependence of magnetization on the applied magnetic field H for a hard magnet. The interpretation of this diagram is as follows. First, this diagram can provide the maximal saturation magnetization value, M_s. Then, it is clear that the application of the external magnetic field does not immediately change the M_s value. The applied H should reach a certain value, H_c, as the material tends to resist this influence. Therefore, namely the value H_c is the measure of particular magnetic material to resist demagnetization. It is called *intrinsic coercivity.* Hard magnetic materials have high values of H_c. Consequently, soft ones have a low resistance to demagnetization.

If one uses magnetic induction B instead of M, the plot shown in Figure 11.5b can be obtained. This plot is convenient because it can help to determine a key point, $|BH|_{max}$, as indicated in the figure. So, one of the central figures of merit for a PM is its so-called *energy product*, $|BH|_{max}$. This quantity is twice the energy stored in the stray field created by the magnet. In simple terms, it can equivalently be understood as energy density (SI units define it in Joules per cubic meter). Another critical property of a PM is, as discussed above, the resistance to demagnetization, H_c. Of course, it is also essential to take into account that the magnetic state of a material largely depends on its temperature and other key variables such as pressure and the applied external magnetic field. It is also noteworthy that a bulk magnet in the permanently magnetized state is metastable; the ground state would be a state that generates no stray field. However, namely $|BH|_{max}$ and H_c are often of primary importance in designing magnets for electrical motors and generators.

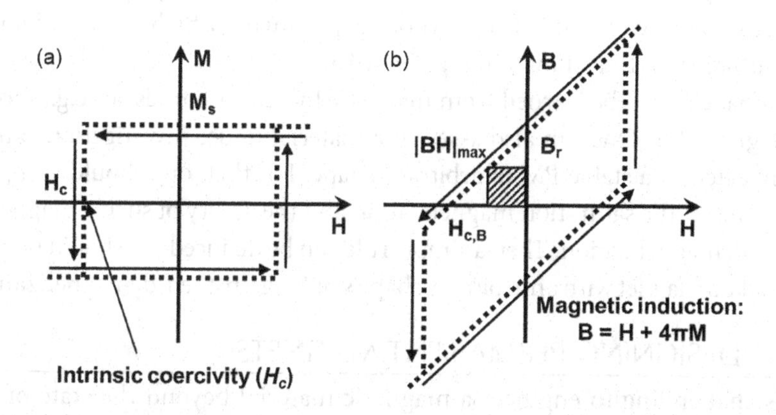

FIGURE 11.5 Ideal hysteresis loops for a PM: (a) an M(H) loop and (b) a B(H) loop. B_r is the residual induction.

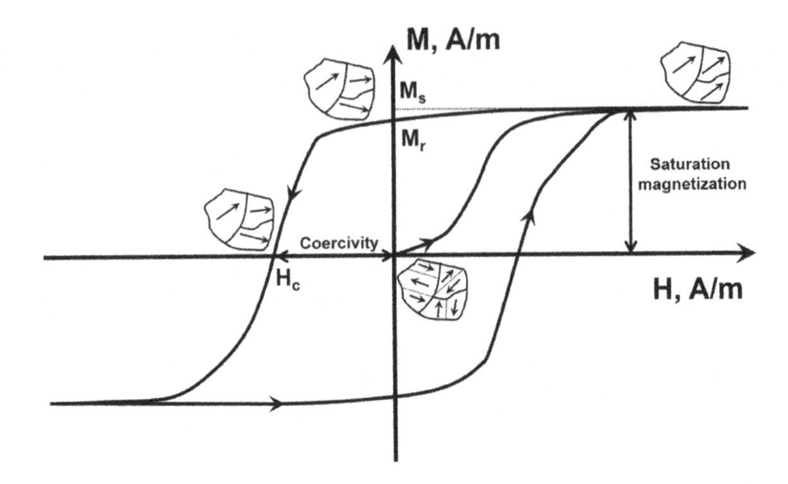

FIGURE 11.6 A realistic hysteresis loop for some hard magnetic material. (Adapted from [11]. M_r is residual magnetization.)

Let us further consider a more realistic M(H) dependence shown in Figure 11.6. This curve starts from a (0,0) coordinate, and total sample magnetization increases as a response to the increasing magnetic field. Magnetization demonstrates a plateau at the saturation value, M_s, where all the domains have magnetization in a uniform direction (see also pictograms in the figure). If one then reverses the direction of the applied field, realistic curves would not usually cross the M-axis at the M_s point but rather below, at M_r, which is a residual magnetization. Moving further, one can observe a sort of smoothed (compared to the corresponding ideal dependence in Figure 11.5a) curve crossing the H-axis at the point H_c, giving the intrinsic coercivity. If one continues, the hysteresis loop can be observed at the end of the whole experiment, which is considerably smoother than that shown in Figure 11.5a.

What else can be learned from magnetic hysteresis curves, as, e.g., shown in Figure 11.6? One can also assess if a material under investigation can be compacted to a stable PM of arbitrary shape. For that, one should compare the value of the saturation magnetization and the ability of such a magnet to resist demagnetization. Then a simple rule can be deduced: H_c should be $> M_r$ to make a magnet with an arbitrary shape stable against self-demagnetization.

11.4 DESIGNING PERMANENT MAGNETS

It is challenging to engineer a magnetic material beyond the state of the art despite having a good fundamental understanding of such systems. However, one can underline several aspects to consider while improving

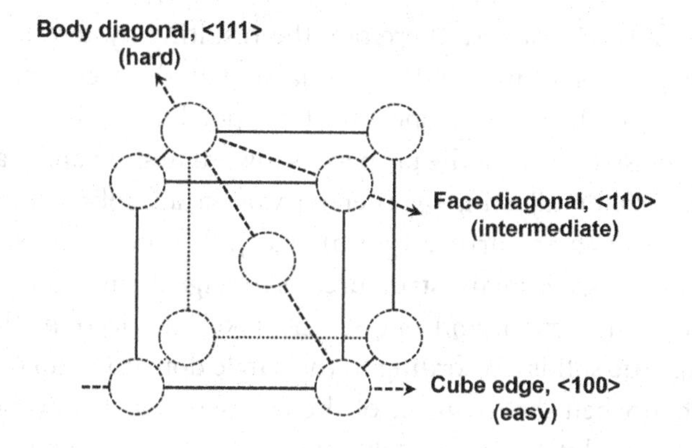

FIGURE 11.7 "Easy", "medium", and "hard" directions of magnetization in a unit cell of bcc iron.

these kinds of materials for energy applications. Besides the choice of basic materials, one of the following "degrees of freedom" to consider is the direction of magnetization (demagnetization) important upon fabrication, characterization, and use [12,13]. There are often "hard" and "soft" directions for magnetization. Figure 11.7 shows a classic example (related to the soft magnetic material, iron, in this case), distinguishing different directions along which magnetization should be dissimilar. These directions are designated as <111>, <100> and <110> in the figure.

Indeed, experimental results (Figure 11.8) demonstrate that along the <111> direction, magnetization is way more complicated compared to

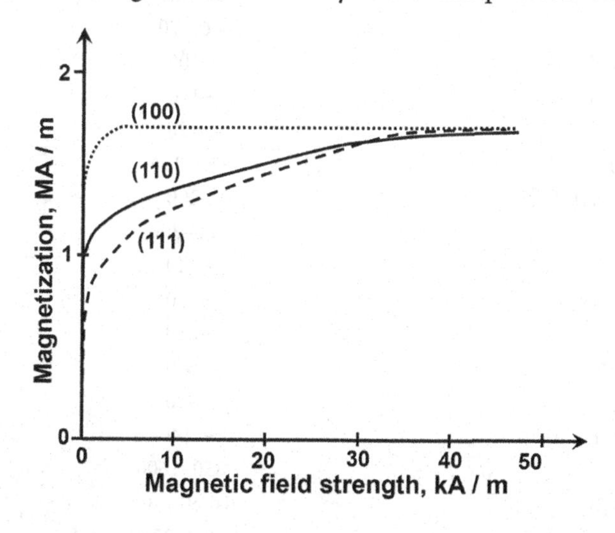

FIGURE 11.8 Magnetization curves for a Fe single crystal with the field, oriented along with the (111), (110), and (100) directions.

the <100> and <110> ones. Therefore, the resulting hysteresis curve will depend on the crystallinity of the material and on the preferential orientation of the crystallites with respect to the applied field. This effect can, in principle, be used to adjust the properties toward desired applications [14].

An approach for adjusting the coercivity of certain solid magnetic materials is to tailor their particle (grain) size [15]. Indeed, considering the magnetic materials' domain structure, ferromagnetism should be a size-dependent phenomenon, and one should take into account the domain size of magnetic solids. According to the single domain theory [16,17], the H_c is optimal when the diameter of the magnetic particles (grains in the polycrystalline solids) is in the stable single-domain size range for various materials. The formation of *multiple* domains in the particles is energetically unfavorable below some *critical particle sizes* (see Table 11.4). The optimum particle size can be reasonably assessed using the following equation:

$$\text{Optimum particle diameter} = 9\sigma_w / 2\pi M_s^2$$

TABLE 11.4 Approximate Single-Domain Critical Particle (Grain) Sizes for Some Magnetic Species

Material	**Critical Particle Sizes, nm. It Can Vary in Different Studies and with Particle Shape**
Fe	~14
FeCo	~52
Ni	~55-85
Co	~60-70
$CoFe_2O_4$	~100
$SmFe_{10}TiV$	~100
Fe_3O_4	~128
γ-Fe_2O_3	~85–166
$Sm_2(Fe,Ti)_{17}N_3$	~200
$Nd_2Fe_{14}B$	~75–300
FePt	~340
$PrCo_5$	~610
CoPt	~610
YCo_5	~680
$CeCo_5$	~920
$Dy_2Fe_{14}B$	~1140
$SmCo_5$	~850–1600
$Tb_2Fe_{14}B$	~1669–1700
$Nd_{16}Fe_{78}Al_{0.4}B_{5.6}$	~1300–2700

Data are from [17,21–27].

where σ_w is the energy per unit surface area associated with the 180° Bloch wall. The latter is a transition region at the boundary between magnetic domains where the magnetization alters from its value in one domain to the other values found in the neighboring one. On the other hand, M_s and M_r also depend on the particle size [18].

In practice, it is often easier to find the optimum experimentally. For that, experimentally measured coercivity is plotted as a function of the average particle (grain) size in the sample. Commonly, a dependence similar to that shown in Figure 11.9 is observed. Initially, when the particle size is very small and the stable single-domain size is not reached, the coercivity also demonstrates very low values. This is because the spins are affected by thermal fluctuations. At some small particle size, the system becomes even superparamagnetic because of this effect. For example, the superparamagnetic size limits for the $Dy_2Fe_{14}B$ and $Tb_2Fe_{14}B$ compounds are as small as approximately 4.7 and 3.9 nm, respectively, at room temperature [19,20]. When the size of the species increases, it passes the maximum corresponding to the optimal domain size or critical size (see Table 11.4). It decreases again upon further particle size increase (multiple domain areas in Figure 11.9).

Increasing the intrinsic coercivity of magnets, for instance, state-of-the-art sintered $Nd_2Fe_{14}B$ solids, is possible even further with specific metal

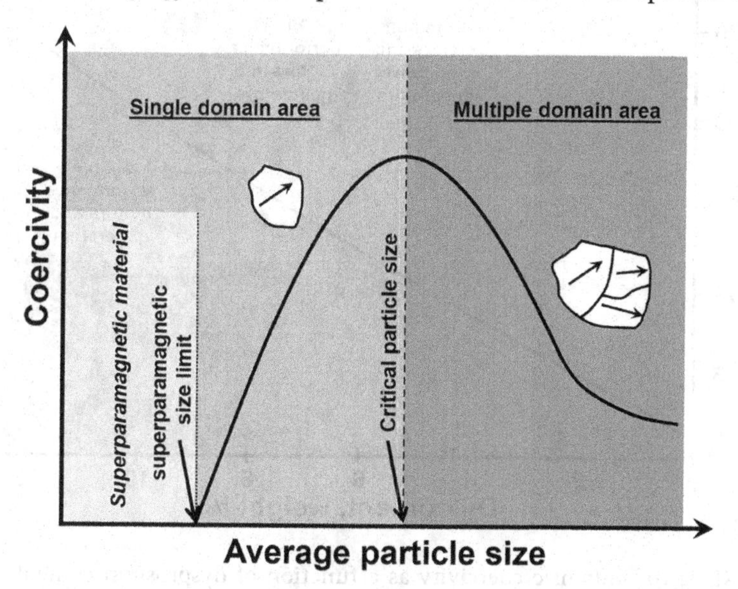

FIGURE 11.9 Approximate change of coercivity of magnetic materials as a function of the particle (grain) size [15,19,28].

ion addition. For that, a certain amount of heavy rare earth ions with high μ_{S+L}, for example, Dy, Tb, or Ho [29–31] (see Table 11.2), is added. Figure 11.10 demonstrates how fast the H_c values grow if just several percent of Dy is added to displace Nd. Just approximately 5% of dysprosium additives are enough to obtain the material suitable for producing common electric motors and wind power generators. If 2%–3% more Dy is added, one can use such magnets to make high and very high-performance devices. It was also particularly remarkable that the best results demonstrated samples composed of $Nd_2Fe_{14}B$ grains with Dy-enriched grain boundaries, surfaces, and subsurfaces [32–35].

If Tb or Ho are added instead of Dy, an even better effect can be obtained [27]. In particular, fewer amounts of additives are needed with the additional benefit of increased operating temperature. However, the availability issues prevent their wider use.

Currently, Dy is only mined at one place in the world (2020) with a limited production, so there might be a risk of exhausting its supply quickly. On the other hand, there are other applications of Tb and Ho. For example, holmium itself is a valuable metal as it is used to create the strongest artificially possible magnetic fields to date. In other words, the availability

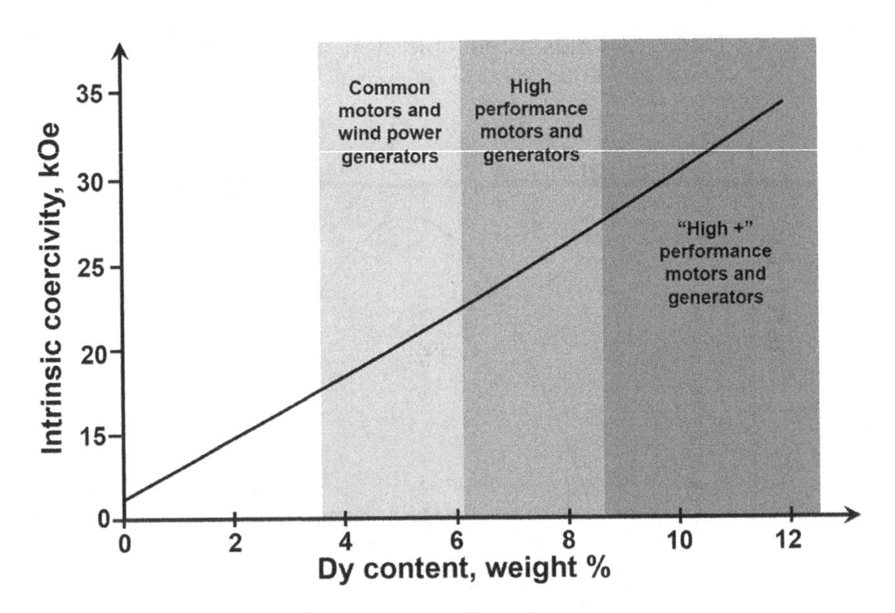

FIGURE 11.10 Intrinsic coercivity as a function of dysprosium content in the neodymium magnets.

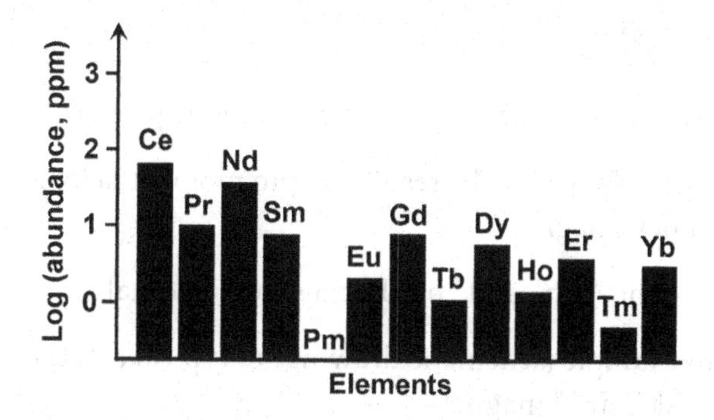

FIGURE 11.11 The abundance of f-elements on Earth in parts per million (ppm).

issue dictates the choice toward Dy to date (see Figure 11.11). Therefore, further rational improvement of methodologies to produce magnetic, ideally rare-earth-free [36], materials with high resistance to demagnetization is of paramount importance.

11.5 SUMMARY AND CONCLUSIONS

Magnetic materials play an essential role in energy applications, particularly in electrical generators and motors used for "green" technologies. The basic theory and methodological aspects related to PMs are relatively well-developed, where the concepts have, in general, high predictive power to discover new materials. However, the development of new generations of affordable magnets meets at least two challenges: the "physical constraints" and availability of rare-earth elements, e.g., dysprosium, Dy. There are at least two important figures of merit for magnetic materials designed for energy applications. These are energy products, which should be as high as possible, and high resistance to self-demagnetization. Nowadays, the state-of-the-art PMs with the highest energy product are neodymium-based magnets like $Nd_2Fe_{14}B$. With the addition of Dy up to 10%–12%, one can significantly increase their resistivity toward self-demagnetization and construct high-performance generators for the wind turbines and motors. The latter are important for electric cars powered by different types of batteries, supercapacitors, or hydrogen fuel. Additional degrees of freedom to design new high-performance PMs include a variation of the composition and material particle size to increase and control coercivity. These approaches should be viable to optimize the use of rare-earth elements and find new, more affordable magnetic materials for energy applications.

11.6 QUESTIONS

1. What are the approaches to predict magnetic properties?
2. How does calculating the resulting spin moment help in the development of new magnets?
3. What are the "soft" and "hard" magnetic materials?
4. Give an example (schematic drawing) of typical M(H) curves for the "soft" and "hard" magnets.
5. Analyze the hysteresis loop shown in Figure 11.6. Can a magnet of arbitrary shape be manufactured from such a material demonstrating significant resistance toward self-demagnetization?
6. Name state-of-the-art magnetic materials with the highest energy product $|BH|_{max}$.
7. Analyze what limits the broader application of PMs as energy materials.
8. Explain what the driving forces responsible for the formation of the ferromagnetic domains are.
9. Analyze why the "neodymium-dysprosium" magnets play the dominant role in, e.g., wind generators.

REFERENCES

1. Hirosawa, S.; Nishino, M.; Miyashita, S. 2017. Perspectives for high-performance permanent magnets: Applications, coercivity, and new materials. *Advances in Natural Sciences: Nanoscience and Nanotechnology* 8:013002.
2. Constantinides, S. 2016. Permanent magnets in a changing world market. *Magnetics Business & Technology*, February 14.
3. Agamloh, E.; von Jouanne, A.; Yokochi, A. 2020. An overview of electric machine trends in modern electric vehicles. *Machines* 8:20.
4. Gutfleisch, O.; Willard M.A.; Brück, E.; Chen C.H.; Sankar, S.G.; Liu, J.P. 2011. Magnetic materials and devices for the 21st centur*Science*y: Stronger, lighter, and more energy efficient. *Advanced Materials* 23:821–842.
5. Went, J.; Rathenau, G.W.; Gorter, E.W.; van Oosterhout, G.W. 1951. Hexagonal iron-oxide materials as permanent magnet materials. *Physical Reviews* 86:424–425.
6. Robinson, A.L. 1984. Powerful new magnet material found. 223:920–922.

7. Lee, S.H.; Oh, H.S.; Kim, H.S. 2017. Experiment research on motor efficiency improvement through the eddy current reduction using segmentation of rotor magnet. *20th International Conference on Electrical Machines and Systems (ICEMS)* 2017:1–4.
8. Rahman, M.A. 2012. History of interior permanent magnet motors. *IEEE Industry Applications Magazine* 19(1):10–15.
9. Frenkel, J.; Doefman, J. 1930. Spontaneous and induced magnetisation in ferromagnetic bodies. *Nature* 126:274–275.
10. Kittel, C. 1946. Theory of the structure of ferromagnetic domains in films and small particles. *Physical Review* 70:965–971.
11. Coey, J.M.D. 2011. Hard magnetic materials: A perspective. *IEEE Transactions on Magnetics* 47:4671–4681.
12. Coey, J.M.D. 2020. Perspective and prospects for rare earth permanent magnets. *Engineering* 6:119–131.
13. Herbst, J.F. 1991. $R_2Fe_{14}B$ materials: Intrinsic properties and technological aspects. *Reviews of Modern Physics* 63:819–898.
14. Fischbacher, J.; Kovacs, A.; Oezelt, H.; Gusenbauer, M.; Suess, D.; Schrefl, T. 2017. *Applied Physics Letters* 11:192407.
15. Shen, B.; Sun, S. 2020. Chemical synthesis of magnetic nanoparticles for permanent magnet applications. *Chemistry - A European Journal* 26: 6757–6766.
16. Kittel, C. 1949. Physical theory of ferromagnetic domains. *Reviews of Modern Physics* 21:541–583.
17. Nothnagel, P.; Müller, K.H.; Eckert, D.; Handstein, A. 1991. The influence of particle size on the coercivity of sintered NdFeB magnets. *Journal of Magnetism and Magnetic Materials* 101:379–381.
18. Li, Q.; Kartikowati, C.W.; Horie, S.; Ogi, T.; Iwaki, T.; Okuyama, K. 2017. Correlation between particle size/domain structure and magnetic properties of highly crystalline Fe_3O_4 nanoparticles. *Scientific Reports* 7:9894.
19. Yue, M.; Liu, R.M.; Liu, W.Q.; Zhang, D.T.; Zhang, J.X.; Guo, Z.H.; Li, W. 2012. Ternary DyFeB nanoparticles and nanoflakes with high coercivity and magnetic anisotropy. *IEEE Transactions on Nanotechnology* 11:651–653.
20. Liu, R.M.; Yue, M.; Liu, W.Q.; Zhang, D.T.; Zhang, J.X.; Guo, Z.H.; Li, W. 2011. Structure and magnetic properties of ternary Tb-Fe-B nanoparticles and nanoflakes. *Applied Physics Letters* 99:162510.
21. Leslie-Pelecky, D.L.; Rieke, R.D. 1996. Magnetic properties of nanostructured materials. *Chemistry of Materials* 8:1770–1783.
22. Castle, E.; Sheridan, R.; Zhou, W.; Grasso, S.; Walton, A.; Reece, M.J. 2017. High coercivity, anisotropic, heavy rare earth-free Nd-Fe-B by flash spark plasma sintering. *Scientific Reports* 7:11134.
23. Girt, Er.; Krishnan, K.M.; Thomas, G.; Girt, E.; Altounian, Z. 2001. Coercivity limits and mechanism in nanocomposite Nd–Fe–B alloys. *Journal of Magnetism and Magnetic Materials* 231:219–230.
24. Livingston, J.D.; McConnell, M.D. 1972. Domain-wall energy in cobalt-rare-earth compounds. *Journal of Applied Physics* 43:4756–4762.

25. Lee, S.K.; Das, B.N.; Harris, V.G. 1999. Magnetic structure of single crystal $Tb_2Fe_{14}B$. *Journal of Magnetism and Magnetic Materials* 207:137–145.
26. Poudyal, N.; Liu, J.P. 2013. Advances in nanostructured permanent magnets research. *Journal of Physics D: Applied Physics* 46:043001.
27. Trinh, T.T.; Kim, J.; Sato, R.; Matsumoto, K.; Teranishi, T. 2021. Synthesis of mesoscopic particles of multi-component rare earth permanentmagnet compounds. *Science and Technology of Advanced Materials* 22:37–54.
28. Tang, H.; Mamakhela, M.A.H.; Christensen, M. 2020. Enhancing the coercivity of $SmCo_5$ magnet through particle size control. *Journal of Materials Chemistry C* 8:2109–2116.
29. Binnemans, K.; Jones, P.T.; Müller, T.; Yurramendi, L. 2018. Rare earths and the balance problem: How to deal with changing markets? *Journal of Sustainable Metallurgy* 4:126–146.
30. Cao, X.; Chen, L.; Guo, S.; Chen, R.; Yan, G.; Yan, A. 2016. Impact of TbF_3 diffusion on coercivity and microstructure in sintered Nd–Fe–B magnets by electrophoretic deposition. *Scripta Materialia* 116:40–43.
31. Soderžnik, M.; Žužek Rožman, K.; Kobe, S.; McGuiness, P. 2012. The grain-boundary diffusion process in Nd–Fe–B sintered magnets based on the electrophoretic deposition of DyF_3. *Intermetallics* 23: 158–162.
32. Sugimoto, S. 2011. Current status and recent topics ofrare-earth permanent magnets. *Journal of Physics D: Applied Physics* 44:064001.
33. Nakamura, H.; Hirota, K.; Shimao, M.; Minowa, T.; Honshima, M. 2005. Magnetic properties of extremely small Nd-Fe-B sintered magnets. *IEEE Transactions on Magnetics* 41:3844–3846.
34. Hirota, K.; Nakamura, H.; Minowa, T.; Honshima, M. 2006. Coercivity enhancement by the grain boundary diffusion process to Nd–Fe–B sintered magnets. *IEEE Transactions on Magnetics* 42:2909–2911.
35. Chen, F. 2020. Recent progress of grain boundary diffusion process of Nd-Fe-B magnets. *Journal of Magnetism and Magnetic Materials* 514:167227.
36. Cui, J.; Kramer, M.; Zhou, L.; Liu, F.; Gabay, A.; Hadjipanayis, G.; Balasubramanian, B.; Sellmyer, D. 2018. Current progress and future challenges in rare-earth-free permanent magnets. *Acta Materialia* 158:118–137.

CHAPTER 12

Materials for Hydrogen Fuel Storage

12.1 MOTIVATION

Hydrogen is the fuel of the universe, and human civilization, again and again, tries to develop technologies to use it widely [1]. The vision of establishing a sort of hydrogen-powered society can be traced back to Poul La Cour, who used H_2 to store energy generated by wind in 1895 with the production of *ca* 1000 L of H_2 per hour, which was then stored in a gas tank [2]. Interestingly, hydrogen has been used for energy provision since the 1800s. It was a major constituent, up to 50 vol.%, of the so-called "city gas", i.e., the fuel gas manufactured using synthetic gas by gasification of coal, wood, or waste. This city gas was used in households in eastern USA and Europe from *ca* 1850 until the middle of the 20th century. The water-gas reaction utilized in the production of the synthetic gas was also one of the first in the attempts to provide energy utilizing hydrogen [3]:

$$C + H_2O(\text{steam}) \rightarrow CO + H_2$$

In the past and today, there were, and there are, however, several challenges to address when using H_2 as a fuel. Apart from the problems of generation and consumption associated with electrolyzers and fuel cells,

DOI: 10.1201/9781003025498-12

which require efficient electrocatalysts, there are issues related to hydrogen storage for stationary, portable, and automotive applications [4–8].

There are presently three major competing technologies for hydrogen storage: compressed gas cylinders, liquid hydrogen tanks, and metal hydrides. The following table (Table 12.1) briefly compares the volumetric hydrogen capacity for these three technologies. As one can see from the table, some solid-state hydrides can be approximately three times more efficient in storing hydrogen, which is particularly important if H_2 is used as a fuel for transportation [9].

Storing hydrogen in compressed gas cylinders is currently a well-established and dominating approach in large-scale systems [10] and cars. For example, the *Toyota Mirai* fuel cell car presently uses two 70 MPa high-pressure hydrogen tanks. Still, improvements in the H_2 storage systems would be desired. For instance, to achieve a driving range of ~500 km for a conventional vehicle with today's diesel technology, it requires a tank system of approximately 43 kg with a volume of ~46 L. A car driven by a polymer electrolyte membrane fuel cell with H_2 will need a 700 bar high-pressure tank system of approximately 125 kg and *ca* 260 L to store enough hydrogen to achieve the same driving range.

Moreover, standard compressed gas cylinder storage and transport systems might experience the embrittlement phenomenon in which various metals, such as high-strength steel, aluminum, and titanium alloys, become brittle and eventually crack under load following exposure to hydrogen. This process can also cause problems for hydrogen steel piping and other components of, e.g., H_2 gas refueling stations. Similar problems, together with additional expenses on maintaining low temperature, are the "satellites" of the liquid hydrogen storage systems. Therefore, alternative solutions are under investigation and testing recently [11]. One alternative, as mentioned above, is to store H_2 in metal hydrides or similar systems [12–17], which are briefly described in the next section.

TABLE 12.1 Comparison of Three Major Competing Technologies for Hydrogen Storage

Storage System	**Volumetric Hydrogen Capacity in kg of H_2 per cubic meter**
Compressed H_2 under 700 bar pressure	~42
Liquid H_2 at −252°C in a tank (1.013 bar)	~71
Solid state hydrides	Up to ~150

12.2 METAL HYDRIDES FOR HYDROGEN STORAGE

Many metals and alloys react reversibly with hydrogen to form metal hydrides according to the following scheme:

$$\text{Metal(alloy)} + x/2\ H_2 \rightarrow \text{Metal(alloy)}H_x + Q$$

Probably the first observation of such kind of phenomenon was made by Thomas Graham in 1866 with respect to metallic palladium [18]. He observed that one unit volume of palladium could hold impressive 643.3 volumes of hydrogen:

$$Pd + H_2 \rightarrow PdH_x$$

In fact, palladium hydride is not a stoichiometric chemical compound but simply a metal in which H_2 is dissolved and stored in the solid-state phase, in the space between the Pd atoms of the crystal lattice of the metal.

Further extensive investigations of metal hydrides were initiated after World War II. These investigations were primarily driven by nuclear reactor applications to understand the hydride-caused embrittlement of reactor metals, such as Zr. Another motivation was to take advantage of the high content of hydrogen in hydrides to scatter or shield from energetic neutrons in nuclear reactors.

Zirconium alloys in the water-cooled reactors were known to absorb hydrogen and precipitate zirconium hydrides upon cooling. Therefore, zirconium–hydrogen systems were studied the most in close reference to the development of nuclear reactors. The first intermetallic hydride reported concerning these studies was $ZrNiH_3$.

As a result of subsequent studies, the five most promising compositions for H_2 storage in metal hydrides were empirically identified. They have the general formula A_mB_n, where A and B are elements with high and low affinities for hydrogen, respectively. Some examples are given in Table 12.2.

TABLE 12.2 Examples of A_mB_n Compositions for H_2 Storage in Metal Hydrides

Compositions	Element A	Element B	Examples
AB	Zr, Ti	Ni	ZrNi, TiNi
AB_2	La, Y	Ni, Mn	$LaNi_2$, YNi_2, YMn_2
A_2B	Mg	Ni, Co	Mg_2Ni, Mg_2Co
AB_3	La, Y	Ni, Co	$LaCo_3$, YNi_3
AB_5	Ca, La	Ni, Cu, Fe	$CaNi_5$, $LaCu_5$, $LaFe_5$

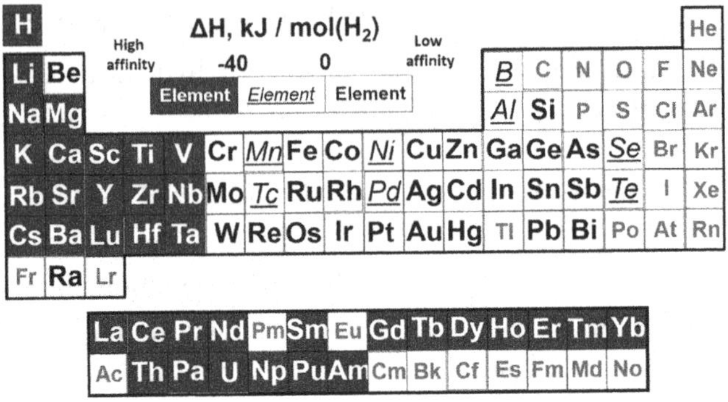

FIGURE 12.1 Elements with relatively high ($\Delta H < -40$ kJ /mol[H_2]) and low ($\Delta H > -40$ kJ/mol[H_2]) affinities to hydrogen

Figure 12.1 highlights the elements in the periodic table with relatively low and high affinities to hydrogen quantified by the heat of formation of respective hydrides.

One should note that many conventional metal hydrides based on metals such as V, Nb, Pd, Li, Na, etc. have gravimetric capacities, which are too low for the intended commercial applications in mobile hydrogen storage except for LiH, which has a high capacity but a very high H_2 desorption temperature. So, what would be an ideal hydride for hydrogen storage? The Department of Energy of the United States of America suggests that the suitable materials for hydrogen storage in hydrides would provide a compromise between physisorption and chemisorption. That means it is in the ΔH range between 10 and 60 kJ/mol[H_2], which is strong enough to hold H_2 but weak enough to allow for rapid desorption [19].

It is possible to assess and predict the heat of formation reasonably. It was found that in many cases, there is a linear relationship between the heat of formation, ΔH, of a metal hydride and the characteristic energy, ΔE, of the electronic band structure of the host metal [20]:

$$\Delta H = \alpha \cdot \Delta E + \beta$$

with $\Delta E = E_F - E_S$ (where E_F is the Fermi energy, and E_S is the center of the lowest conduction band of the host metal, $\alpha \sim 59.24$ kJ [eV mol H_2]$^{-1}$ and $\beta \sim -270$ kJ [mol H_2]$^{-1}$ and ΔE in eV.

Since a high weight fraction of H_2 in the solid-state hydrogen storage materials is necessary for practical transportation applications, research to enhance the hydrogen capacity in metal hydrides has been recently focused on those based on lightweight elements such as lithium (Li), magnesium (Mg), or aluminum (Al) [21–25]. Magnesium hydride is among the most important and most comprehensively investigated light hydrides [26]. MgH_2 itself has a relatively high reversible storage capacity, approximately 7.6 wt.%. Furthermore, Mg is the 8th most abundant element on the Earth and thus comparably affordable. Its potential usage initially was hindered because of its rather sluggish dehydrogenation dynamics and relatively unfavorable reaction enthalpies.

Figure 12.2 compares some hydrides, which are currently under investigation for reversible hydrogen adsorption and desorption. It should be noted that besides high volumetric and gravimetric capacity, these hydrides should maintain their reactivity and capacity over thousands of cycles for practical applications [27].

In order to facilitate hydrogen absorption, the following steps, which generally occur sequentially, should be optimized toward lowering the corresponding energy barriers (Figure 12.3).

1. H_2 physisorption at the surface of the storage material
2. Subsequent hydrogen chemisorption
3. Hydrogen penetration into the subsurface
4. Diffusion of H_2 into bulk
5. Nucleation and growth of the hydride phase

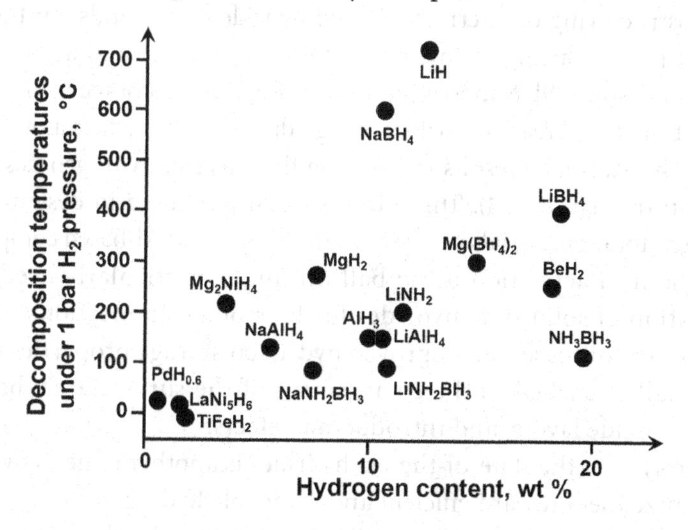

FIGURE 12.2 Overview of various solid-state hydrides. (Adapted from [19].)

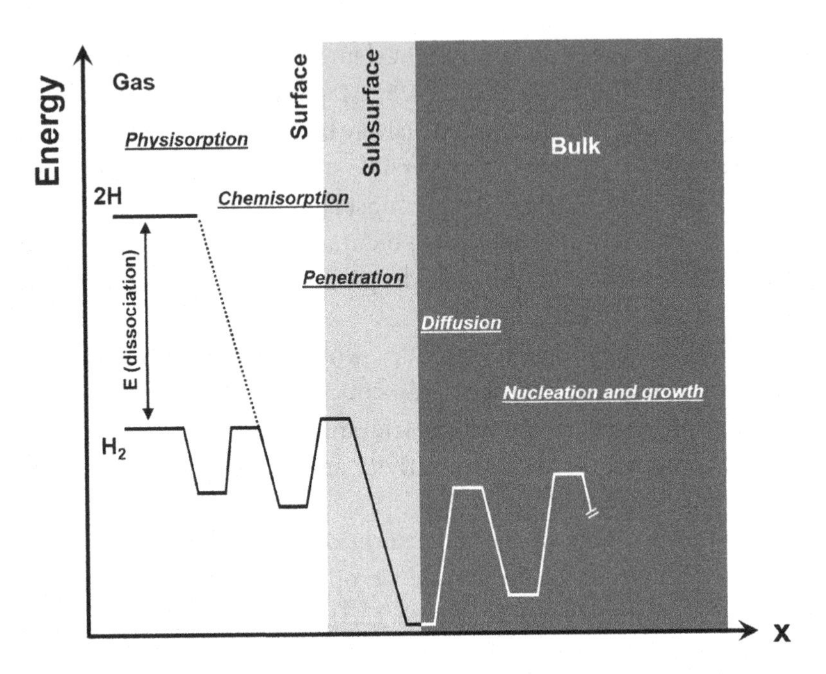

FIGURE 12.3 Schematic energy diagram of energy barriers faced by H_2 during absorption/desorption in/from a metal.

The initial stages of H_2 absorption can be facilitated by immobilizing substances promoting, e.g., hydrogen spillover by a catalyst system. For that, one can use, for instance, submonolayers or small clusters of metallic palladium [28].

Nanostructuring of intermetallic and hydride compounds, on the other hand, helps in reducing the hydrogen diffusion distance and increasing the H_2 exchange rate [29]. Nanoscale particles of hydrogen storage materials also demonstrate an increase in surface energy due to lower surface atom coordination. This normally results in lowering the effective hydrogen desorption temperature (Figure 12.4). This effect is extensively used in designing new hydrogen storage materials using well-known Mg- and Al-based compounds.

Mechanical activation using ball milling is particularly used in the preparation of solid state hydrides for H_2 storage applications (see, e.g. [30–32]). In this case, the improved hydrogen storage properties may be additionally ascribed to the enlargement of the surface area, breaking the outer oxide layers, and introducing defects.

Hydrolysis of the state-of-the-art hydrides is another issue to overcome to optimize them toward efficient and reversible hydrogen storage. This is because the hydrogen gas, especially that obtained by the electrolysis of

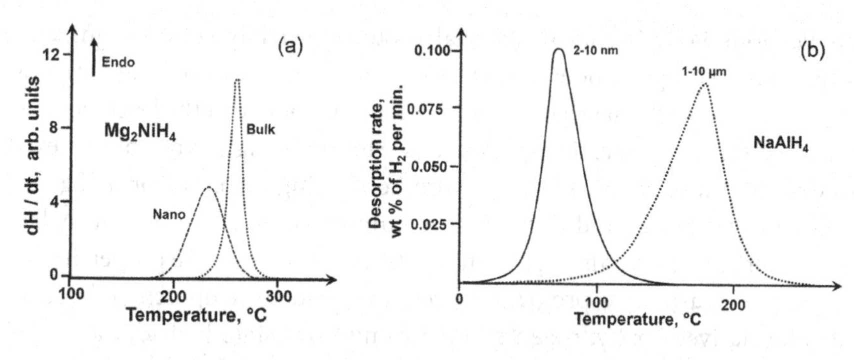

FIGURE 12.4 Desorption of hydrogen from bulk and nanostructured (a) Mg_2NiH_4 (Adapted from [29]) and (b) $NaAlH_4$ (Adapted from [34].)

water, often contains a small amount of H_2O molecules. The hydrolysis takes place according to the following schemes, resulting in hydroxides, which block the surface of the particles.

$$MgH_2 + 2H_2O \rightarrow 2H_2 + Mg(OH)_2$$

$$LiAlH_4 + 4H_2O \rightarrow LiOH + Al(OH)_3 + 4H_2$$

This issue can be solved if species of the hydrogen storage materials are encapsulated by a selectively gas permeable polymer, for example, poly(methyl methacrylate). The polymer matrix protects the particles from water molecules, and, at the same time, it is permeable to smaller hydrogen molecules that can reach the surface of the species [33].

Nonmetallic materials for hydrogen storage are also under extensive experimental and theoretical investigation [35]. Among them are carbon-based materials like graphene and its composites [36–40], C-nanotubes [41], or metal-organic frameworks [42–46]. The idea, in this case, is to use the ability of the hydrogen gas to be adsorbed in arrays of carbon nanotubes or similar structures. Another alternative is to explore the so-called clathrates, which are a sort of inclusion compounds in which the guest molecule such as hydrogen is captured in a cage formed by the host molecule or by a lattice of host molecules (e.g., water).

12.3 SUMMARY AND CONCLUSIONS

Storing hydrogen in solid materials is based on the observation that some metals can reversibly absorb hydrogen. However, the practical application of such a finding is rather challenging, especially for automotive applications [47]. The ideal material should reversibly store a significant amount of hydrogen under moderate conditions of pressures and temperatures. To date, such material has not yet been elaborated at the level required for large-scale commercial applications. However, high expectations exist to develop those based on magnesium-containing materials or using carbon nanostructures and clathrates. A number of new lightweight hydrogen-absorbing materials have been identified with attractive properties.

Despite reasonable progress in research and development related to the alternatives for hydrogen storage using solid state hydrides and carbon materials, only a few companies have commercialized these systems by now at a smaller scale (see, e.g., [48]). Most likely, general progress in the generation and consumption of hydrogen as a fuel will stimulate the broader application of such systems in the near future.

12.4 QUESTIONS

1. What is the major benefit of metal hydride compounds to use for hydrogen storage? Analyze the advantages and disadvantages.
2. What are the typical bond types in metal hydrides?
3. What are the optimum binding energies in metal hydrides for H_2 storage?
4. What are the main steps in the process of hydrogen absorption?
5. What is the impact of binding energies in metal hydrides on the reversibility of hydrogenation/dehydrogenation?
6. What are the typical strategies to improve metal hydride materials for hydrogen storage?
7. Name state-of-the-art and alternative nonmetallic materials for hydrogen storage.

REFERENCES

1. Pellow, M.A.; Emmott, C.J.M.; Barrnhart, C.J.; Benson, S.M. 2015. Hydrogen or batteries for grid storage? A net energy analysis. *Energy & Environmental Science* 8:1938–1952.
2. Ley, M.B.; Jepsen, L.H.; Lee, Y.-S.; Cho, Y.W.; Bellosta von Colbe, J.M.; Dornheim, M.; Rokni, M.; Jensen, J.O.; Sloth, M.; Filinchuk, Y.; Jørgensen, J.E.; Besenbacher, F.; Jensen, T.R. 2014. Complex hydrides for hydrogen storage – new perspectives. *Materials Today* 17:122–128.

3. Varin, R.A.; Czujko, T.; Wronski, Z.S. 2009. *Nanomaterials for Solid State Hydrogen Storage*. Springer: Berlin.
4. Andersson, J.; Grönkvist, S. 2019. Large-scale storage of hydrogen. *International Journal of Hydrogen Energy* 44:11901–11919.
5. Eberle, U.; Felderhoff, M.; Schüth, F.M. 2009. Chemical and physical solutions for hydrogen storage. *Angewandte Chemie International Edition* 48:6608–6630.
6. Preuster, P.; Alekseev, A.; Wasserscheid, P. 2017. Hydrogen storage technologies for future energy systems. *Annual Review of Chemical and Biomolecular Engineering* 8:445–471.
7. Wolf, E. 2015. Large-scale hydrogen energy storage. In: *Electrochemical Energy Storage for Renewable Sources and Grid Balancing*. Editors: Garche, J. Elsevier: Amsterdam, pp. 129–142.
8. Niaz, S.; Manzoor, T.; Pandith, A.H. 2015. Hydrogen storage: Materials, methods and perspectives. *Renewable and Sustainable Energy Reviews* 50:457–469.
9. Hwang, H.T.; Varma, A. 2014. Hydrogen storage for fuel cell vehicles. *Current Opinion in Chemical Engineering* 5:42–48.
10. Elberry, A.M.; Thakur, J.; Santasalo-Aarnio, A.; Larmi, M. 2021. Large-scale compressed hydrogen storage as part of renewable electricity storage systems. *International Journal of Hydrogen Energy* 46:15671–15690.
11. Rusman, N.A.A.; Dahari, M. 2016. A review on the current progress of metal hydrides materials for solid state hydrogen storage applications. *International Journal of Hydrogen Energy* 41:12108–12126.
12. Łodziana, Z.; Dębski, A.; Cios, G.; Budziak, A. 2019. Ternary $LaNi_{4.75}M_{0.25}$ hydrogen storage alloys: surface segregation, hydrogen sorption and thermodynamic stability. *International Journal of Hydrogen Energy* 44:1760–1773.
13. Jena, P. 2011. Materials for hydrogen storage: Past, present, and future. *Journal of Physical Chemistry Letters* 2:206–211.
14. Orimo, S.-I.; Nakamori, Y.; Eliseo, J.R.; Züttel, A.; Jensen, C.M. 2007. Complex hydrides for hydrogen storage. *Chemical Reviews* 107:4111–4132.
15. Züttel, A. 2003. Materials for hydrogen storage. *Materials Today* 6:24–33.
16. Lototskyy, M.V.; Yartys, V.A.; Pollet, B.G.; Bowman, R.C. 2014. Metal hydride hydrogen compressors: a review. *International Journal of Hydrogen Energy* 39:5818–5851.
17. Morris, L.; Hales, J.J.; Trudeau, C.; Georgiev, P.; Embs, J.P.; Eckert, J.; Kaltsoyannis, N.; Antonelli, D.M. 2019. A manganese hydride molecular sieve for practical hydrogen storage under ambient conditions. *Energy & Environmental Science* 12:1580–1591.
18. Graham, T. 1866. On the absorption and dialytic separation of gases by colloid septa. *Philosophical Transactions of the Royal Society (London)* 156:399–439.
19. Yi, J.; Sun, C.; Shen, S.; Zou, J.; Mao, S.S.; Yao, X. 2015. Combination of nanosizing and interfacial effect: Future perspective for designing Mg-based nanomaterials for hydrogen storage. *Renewable and Sustainable Energy Reviews* 44:289–303.

20. Griessen, R.; Driessen, A. 1984. Heat of formation and band structure of binary and ternary metal hydrides. *Physical Review B* 30:4372.
21. Shao, H.; Xin, G.; Zheng, J.; Li, X.; Akiba, E. 2012. Nanotechnology in Mg-based materials for hydrogen storage. *Nano Energy* 1:590–601.
22. Shao, H. (Ed.). 2018. *Hydrogen Storage: Preparation, Applications and Technology.* Nova Science Publishers: Hauppauge, NY, 290 p.
23. de Jongh, P.E.; Allendorf, M.; Vajo, J.J.; Zlotea, C. 2013. Nanoconfined light metal hydrides for reversible hydrogen storage. *MRS Bulletin* 38:488–494.
24. Li, Q.; Qiu, S.; Wu, C.; Lau, K.T.; Sun, C.; Jia, B. 2021. Computational investigation of MgH_2/graphene heterojunctions for hydrogen storage. *Journal of Physical Chemistry C* 125:2357–2363.
25. Samolia, M.; Dhilip Kumar, T.J. 2014. Fundamental studies of H_2 interaction with MAl_3 clusters [M=Li, Sc, Ti, Zr]. *Journal of Alloys and Compounds* 588:144–152.
26. Wang, Y.; Chen, X.; Zhang, H.; Xia, G.; Sun, D.; Yu, X. 2020. Heterostructures built in metal hydrides for advanced hydrogen storage reversibility. *Advanced Materials* 32:2002647.
27. Boateng, E.; Chen, A. 2020. Recent advances in nanomaterial-based solid-state hydrogen storage. *Materials Today Advances* 6:100022.
28. Adams, B.D.; Chen, A. 2011. The role of palladium in a hydrogen economy. *Materials Today* 14:282–289.
29. Chen, X.; Li, C.; Grätzel, M.; Kostecki, R.; Mao, S.S. 2012. Nanomaterials for renewable energy production and storage. *Chemical Society Reviews* 41:7909–7937.
30. Huot, J.; Ravnsbæk, D.B.; Zhang, J.; Cuevas, F.; Latroche, M.; Jensen, T.R. 2013. Mechanochemical synthesis of hydrogen storage materials. *Progress in Materials Science* 58:30–75.
31. Sulaiman, N.N.; Ismail, M.; Rashid, A.H.A.; Ali, N.A.; Sazelee, N.A.; Timmiati, S.N. 2021. Hydrogen storage properties of Mg-Li-Al composite system doped with Al_2TiO_5 catalyst for solid-state hydrogen storage. *Journal of Alloys and Compounds* 870:159469.
32. Milanović, I.; Milošević, S.; Matović, L.; Vujasin, R.; Novaković, N.; Checchetto, R.; Grbović Novaković, J. 2013. Hydrogen desorption properties of MgH_2/$LiAlH_4$ composites. *International Journal of Hydrogen Energy* 38:12152–12158.
33. Jeon, K.J.; Moon, H.R.; Ruminski, A.M.; Jiang, B.; Kisielowski, C.; Bardhan, R.; Urban, J.J. 2011. Air-stable magnesium nanocomposites provide rapid and high-capacity hydrogen storage without using heavy-metal catalysts. *Nature Materials* 10:286–290.
34. Singh, R.; Altaee, A.; Gautam, S. 2020. Nanomaterials in the advancement of hydrogen energy storage. *Heliyon* 6:e04487.
35. Liu, S.; Liu, J.; Liu, X.; Shang, J.; Xu, L.; Yu, R.; Shui, J. 2021. Hydrogen storage in incompletely etched multilayer Ti_2CT_x at room temperature. *Nature Nanotechnology* 16:331–336.

36. Nagar, R.; Vinayan, B.P.; Samantaray, S.S.; Ramaprabhu, S. 2017. Recent advances in hydrogen storage using catalytically and chemically modified graphene nanocomposites. *Journal of Materials Chemistry A* 5:22897–22912.
37. Liu, Y.; Zhang, Z.; Wang, T. 2018. Enhanced hydrogen storage performance of three-dimensional hierarchical porous graphene with nickel nanoparticles. *International Journal of Hydrogen Energy* 43:11120–11131.
38. Cui, H.; Tian, W.; Zhang, Y.; Liu, T.; Wang, Y.; Shan, P.; Chen, Y.; Yuan, H. 2020. Study on the hydrogen storage performance of graphene(N)–Sc–graphene(N) structure. *International Journal of Hydrogen Energy* 45: 33789–33797.
39. Guo, J.H.; Li, S.J.; Su, Y.; Chen, G. 2020. Theoretical study of hydrogen storage by spillover on porous carbon materials. *International Journal of Hydrogen Energy* 45:25900–25911.
40. Singh, S.B.; De, M. 2021. Effects of gaseous environments on physicochemical properties of thermally exfoliated graphene oxides for hydrogen storage: A comparative study. *Journal of Porous Materials* 28:875–888.
41. Jastrzębski, K.; Kula, P. 2021. Emerging technology for a green, sustainable energy promising materials for hydrogen storage, from nanotubes to graphene—a review. *Materials* 14:2499.
42. Sule, R.; Mishra, A.K.; Nkambule, T.T. 2021. Recent advancement in consolidation of MOFs as absorbents for hydrogen storage. *International Journal of Energy Research* 45:12481–12499.
43. Ren, J.; North, B.C. 2014. Shaping porous materials for hydrogen storage applications: A review. *Journal of Technology Innovations in Renewable Energy* 3:12–20.
44. Rösler, C.; Fischer, R.A. 2015. Metal–organic frameworks as hosts for nanoparticles. *CrystEngComm* 17:199–217.
45. Ahmed, A.; Seth, S.; Purewal, J.; Wong-Foy, A.G.; Veenstra, M.; Matzger, A.J.; Siegel, D.J. 2019. Exceptional hydrogen storage achieved by screening nearly half a million metal-organic frameworks. *Nature Communications* 10:1568.
46. Goldsmith, J.; Wong-Foy, A.G.; Cafarella, M.J.; Siegel, D.J. 2013. Theoretical limits of hydrogen storage in metal–organic frameworks: opportunities and trade-offs. *Chemistry of Materials* 25:3373–3382.
47. Durbin, D.; Malardier-Jugroot, C. 2013. Review of hydrogen storage techniques for on-board vehicle applications. *International Journal of Hydrogen Energy* 38:14595–14617.
48. HYDROSTIK PRO. https://amtekcompany.com/doc/Horizon/LWH22-10L-5-datasheet.pdf (Accessed: June 2021).

Index

Printed in the United States
by Baker & Taylor Publisher Services

Printed in the United States
by Baker & Taylor Publisher Services